Die Grundlehren der mathematischen Wissenschaften

in Einzeldarstellungen
mit besonderer Berücksichtigung
der Anwendungsgebiete

Band 173

Herausgegeben von

J. L. Doob · A. Grothendieck · E. Heinz · F. Hirzebruch
E. Hopf · H. Hopf · W. Maak · S. MacLane · W. Magnus
M. M. Postnikov · F. K. Schmidt · D. S. Scott · K. Stein

Geschäftsführende Herausgeber

B. Eckmann und B. L. van der Waerden

F. Maeda · S. Maeda

Theory of Symmetric Lattices

Springer-Verlag Berlin Heidelberg New York 1970

Prof. Dr. Fumitomo Maeda†
Late Professor of Mathematics, Hiroshima University

Prof. Dr. Shûichirô Maeda
Professor of Mathematics, Ehime University

Geschäftsführende Herausgeber:

Prof. Dr. B. Eckmann
Eidgenössische Technische Hochschule Zürich

Prof. Dr. B. L. van der Waerden
Mathematisches Institut der Universität Zürich

AMS Subject Classifications (1970):
Primary 06A30 · Secondary 50D05, 46E50, 46L10

ISBN-13: 978-3-642-46250-4 e-ISBN-13: 978-3-642-46248-1
DOI: 10.1007/978-3-642-46248-1

Softcover reprint of the hardcover 1st edition 1970

Title No. 5156

Preface

Of central importance in this book is the concept of modularity in lattices. A lattice is said to be modular if every pair of its elements is a modular pair. The properties of modular lattices have been carefully investigated by numerous mathematicians, including J. von Neumann who introduced the important study of continuous geometry. Continuous geometry is a generalization of projective geometry; the latter is atomistic and discrete dimensional while the former may include a continuous dimensional part.

Meanwhile there are many non-modular lattices. Among these there exist some lattices wherein modularity is symmetric, that is, if a pair (a,b) is modular then so is (b,a). These lattices are said to be M-symmetric, and their study forms an extension of the theory of modular lattices.

An important example of an M-symmetric lattice arises from affine geometry. Here the lattice of affine sets is upper continuous, atomistic, and has the covering property. Such a lattice, called a matroid lattice, can be shown to be M-symmetric. We have a deep theory of parallelism in an affine matroid lattice, a special kind of matroid lattice. Furthermore we can show that this lattice has a modular extension.

On the other hand, an M-symmetric lattice with a modular extension was introduced by R. L. Wilcox, and it bears his name. An affine matroid lattice is an atomistic Wilcox lattice. In general Wilcox lattice, we introduce the concept of point-free parallelism to extend the theory of parallelism in the atomistic case. We may say that matroid lattices and Wilcox lattices are geometric lattices.

The other important examples of M-symmetric lattices appear in functional analysis. One of them is the lattice of closed subspaces of a Hilbert space. More generally the lattice L of closed subspaces of a locally convex space has the following property: Both L and its dual are atomistic and have the covering property. We call such a lattice a DAC-lattice, and we can show that any DAC-lattice is M-symmetric. There is no parallelism in a DAC-lattice, but we find some interesting structural features which are due to the duality of this lattice.

A non-atomistic generalization of a DAC-lattice is, for instance, the projection lattice of a von Neumann algebra which was proved to be M-symmetric recently. Moreover, we can find an example of an M-symmetric lattice with duality which arises from the dimension theory of lattices. We may say that M-symmetric lattices with duality are analytic lattices.

This book consists of eight chapters. The following table shows the contents of these chapters.

	atomistic case	general case
Preliminary and general arguments	Chap. II	Chap. I
Theory of geometric symmetric lattices	Chap. III Chap. IV	Chap. V
Theory of analytic symmetric lattices	Chap. VI Chap. VII	Chap. VIII

Original notes on theory of symmetric lattices, which mainly consist of arguments on geometric symmetric lattices, were written by one of the authors, Fumitomo Maeda. After his death (in 1965), these notes were enlarged by Shûichirô Maeda. In 1967—68, he gave a lecture on theory of symmetric lattices at University of Massachusetts, and the manuscript of this book has been completed after this lecture.

Acknowledgements

I am very much indebted to Professor G. Birkhoff and Professor S. S. Holland, Jr. for their encouragements in the preparation of this book and for their intermediations to the publisher.

I am also indebted to Professor W. J. Strother and D. J. Foulis in Department of Mathematics, University of Massachusetts, who gave me an opportunity of giving a lecture on theory of symmetric lattices and gave me good circumstances of studying this theory.

During my stay at University of Massachusetts, discussions with Professor S. S. Holland, Jr., M. F. Janowitz, and E. A. Schreiner helped inspire much of my work. I wish to express my deep appreciation to them. Finally, I wish to thank R. J. Weaver for his assistance during my lecture.

Matsuyama, in October 1970 S. Maeda

Contents

Chapter I

Symmetric Lattices and Basic Properties of Lattices

Chapter II

Atomistic Lattices and the Covering Property

Chapter III

Matroid Lattices

Chapter IV

Parallelism in Symmetric Lattices

Chapter V

Point-free Parallelism in Symmetric Lattices

Chapter VI

Atomistic Symmetric Lattices with Duality

Chapter VII

Atomistic Lattices of Subspaces of Vector Spaces

Chapter VIII

Orthomodular Symmetric Lattices

Chapter I

Symmetric Lattices and Basic Properties of Lattices

1. Modularity in Lattices

A lattice L is a partially ordered set any two of whose elements a and b have a least upper bound $a \vee b$ and a greatest lower bound $a \wedge b$, which are respectively called the join and the meet of a and b. The least element and the greatest element, if they exist, are denoted by 0 and 1 respectively.

The dual L^* of a lattice L is the lattice defined by the converse order-relation on the same elements.

If L is a lattice, then a sublattice of L is a subset of L which is itself a lattice under the operations of join and meet which occur in L.

When $a<b$ in a lattice L, then the interval $\{x \in L; a \leqq x \leqq b\}$ is a sublattice of L. We denote it by $L[a,b]$. Its dual coincides with $L^*[b,a]$. The set $\{x \in L; a \leqq x\}$ is denoted by $L[a, \rightarrow]$.

Definition (1.1). Let a and b be elements of a lattice L. We say that (a,b) is a *modular pair*, and we write $(a,b)M$, when

$$(c \vee a) \wedge b = c \vee (a \wedge b) \quad \text{for every } c \leqq b.$$

We say that (a,b) is a *dual-modular pair*, and we write $(a,b)M^*$ when

$$(c \wedge a) \vee b = c \wedge (a \vee b) \quad \text{for every } c \geqq b.$$

Evidently, if $a \leqq b$ then $(a,b)M$, $(b,a)M$, $(a,b)M^*$ and $(b,a)M^*$ hold. Note that $(a,b)M^*$ in L if and only if $(a,b)M$ in L^*.

We write $(a,b)\bar{M}$ when the pair (a,b) is not modular.

Lemma (1.2). *Let a be an element of a lattice L. Then, $(a,x)M$ for all $x \in L$ if and only if $(a,x)M^*$ for all $x \in L$.*

Proof. Evidently, $(a,x)M$ for all $x \in L$ if and only if

(1) $\quad (y \vee a) \wedge x = y \vee (a \wedge x) \quad$ for any pair (x,y) with $x \geqq y$.

On the other hand, $(a,x)M^*$ for all $x \in L$ if and only if

(2) $\quad (y \wedge a) \vee x = y \wedge (a \vee x) \quad$ for any pair (x,y) with $x \leqq y$.

It is evident that (1) and (2) are equivalent. □

Lemma (1.3). *Let a and b be elements of a lattice L. If both $(a,b)M$ and $(b,a)M^*$ hold, then the sublattices $L[a,a\vee b]$ and $L[a\wedge b,b]$ are isomorphic by the following mutually inverse mappings: $x\to x\wedge b$ and $y\to y\vee a$.*

Proof. If $x\in L[a,a\vee b]$ then since $(b,a)M^*$ we have

$$(x\wedge b)\vee a=x\wedge(b\vee a)=x.$$

If $y\in L[a\wedge b,b]$ then since $(a,b)M$ we have

$$(y\vee a)\wedge b=y\vee(a\wedge b)=y.$$

Hence the two mappings $x\to x\wedge b$ and $y\to y\vee a$ are mutually inverse, and obviously they are order-preserving. Therefore $L[a,a\vee b]$ and $L[a\wedge b,b]$ are isomorphic. ▯

Lemma (1.4). *Let a and b be elements of a lattice L. If a and b belong to an interval $L[c,d]$ and $(a,b)M$ holds in $L[c,d]$ (in particular, if $(a,b)M$ holds in $L[a\wedge b,a\vee b]$), then $(a,b)M$ holds in L. The same statement on dual-modularity also holds.*

Proof. If $x\leqq b$ in L, then since $x\vee c\in L[c,d]$ and $x\vee c\leqq b$, by the assumption we have

$$(x\vee a)\wedge b=(x\vee c\vee a)\wedge b=(x\vee c)\vee(a\wedge b)=x\vee(a\wedge b).$$

Hence $(a,b)M$ holds in L. ▯

Lemma (1.5). *Let a, b and c be elements of a lattice L.*

(1.5.1) *If $(a,b)M$ and $(a\wedge b,c)M$ then $(a_1,b\wedge c)M$ for any $a_1\in L[a\wedge c,a]$.*
(1.5.2) *If $(a,b)M$ then $(a_1,b_1)M$ for any $a_1\in L[a\wedge b,a]$ and $b_1\in L[a\wedge b,b]$.*
(1.5.3) *If $(a,b)M$ and $a\wedge b=0$ then $(a_1,b_1)M$ for any $a_1\leqq a$ and $b_1\leqq b$.*

Proof. (I) Let $(a,b)M$, $(a\wedge b,c)M$ and $a\wedge c\leqq a_1\leqq a$. If $d\leqq b\wedge c$, then since $d\leqq b$ and $d\leqq c$, we have

$$\begin{aligned}(d\vee a_1)\wedge(b\wedge c)&\leqq(d\vee a)\wedge b\wedge c=\{d\vee(a\wedge b)\}\wedge c=d\vee(a\wedge b\wedge c)\\&=d\vee(a_1\wedge b\wedge c)\leqq(d\vee a_1)\wedge(b\wedge c).\end{aligned}$$

Hence $(a_1,b\wedge c)M$ holds.

(II) Let $(a,b)M$ and $a\wedge b\leqq b_1\leqq b$. Since $(a\wedge b,b_1)M$, by (I) we have $(a_1,b_1)M$ for any $a_1\in L[a\wedge b_1,a]=L[a\wedge b,a]$.

(III) (1.5.3) follows from (1.5.2). ▯

Lemma (1.6). *If $(a,b)M$, $(c,a\vee b)M$ and $c\wedge(a\vee b)\leqq a$, then $(c\vee a,b)M$ and $(c\vee a)\wedge b=a\wedge b$.*

Proof. Let $(a,b)M$, $(c,a\vee b)M$ and $c\wedge(a\vee b)\leqq a$. If $d\leqq b$, then

$$\begin{aligned}(d\vee c\vee a)\wedge b&=\{(a\vee d)\vee c\}\wedge(a\vee b)\wedge b=[(a\vee d)\vee\{c\wedge(a\vee b)\}]\wedge b\\&=(a\vee d)\wedge b=d\vee(a\wedge b)\leqq d\vee\{(c\vee a)\wedge b\}\leqq(d\vee c\vee a)\wedge b.\end{aligned}$$

Hence $(c \vee a, b)M$ holds. Moreover

$$(c \vee a) \wedge b = (a \vee c) \wedge (a \vee b) \wedge b = [a \vee \{c \wedge (a \vee b)\}] \wedge b = a \wedge b. \quad \square$$

Definition (1.7). A lattice L is called *modular* when $(a,b)M$ hold for all elements a and b of L. By (1.2), in a modular lattice, $(a,b)M^*$ hold for all a and b.

A lattice L is called *M-symmetric* (resp. *M*-symmetric*) when $(a,b)M$ implies $(b,a)M$ (resp. $(a,b)M^*$ implies $(b,a)M^*$) in L. Evidently, L is M*-symmetric if and only if L^* is M-symmetric. (Sometimes an M-symmetric lattice is called a *semi-modular* lattice. See Birkhoff [1], p. 83.)

Remark (1.8). It is evident that any sublattice of a modular lattice is modular. While, it follows from (1.4) that if $a<b$ in an M-symmetric (resp. M*-symmetric) lattice L then the interval $L[a,b]$ is M-symmetric (resp. M*-symmetric).

Theorem (1.9). *A lattice L is M-symmetric if in L*

(1.9.1) $$(a,b)M \quad \textit{implies } (b,a)M^*.$$

L is M-symmetric if in L*

(1.9.2) $$(a,b)M^* \quad \textit{implies } (b,a)M.$$

Proof. Let $(a,b)M$ hold in a lattice L satisfying (1.9.1). If c is an element of L such that $a \wedge b \leqq c \leqq a$, then $(c,b)M$ holds by (1.5.2). Since $(b,c)M^*$ holds by (1.9.1), we have

$$(c \vee b) \wedge a = a \wedge (b \vee c) = (a \wedge b) \vee c = c \vee (b \wedge a).$$

Therefore $(b,a)M$ holds in $L[a \wedge b, a \vee b]$, and then by (1.4) $(b,a)M$ holds in L. Thus L is M-symmetric.

The second statement can be proved similarly. $\square$

A lattice L satisfying (1.9.1) (resp. (1.9.2)) may be called *cross-symmetric* (resp. *dual cross-symmetric*).

Definition (1.10). A lattice L with 0 is called *weakly modular* when in L

(1.10.1) $$a \wedge b \neq 0 \quad \text{implies } (a,b)M.$$

It follows from (1.4) that a lattice L with 0 is weakly modular if and only if the sublattice $L[a, \rightarrow]$ is modular for every $a>0$.

Definition (1.11). A lattice L with 0 is called *$\perp$-symmetric* when in L

(1.11.1) $$(a,b)M \text{ and } a \wedge b = 0 \text{ together imply } (b,a)M.$$

It is evident that an M-symmetric lattice with 0 is $\perp$-symmetric and that a weakly modular $\perp$-symmetric lattice is M-symmetric.

Remark (1.12). If $0<a$ in a $\perp$-symmetric lattice L then evidently the interval $L[0,a]$ is $\perp$-symmetric. By the same way as (1.9), we can prove that a lattice L with 0 is $\perp$-symmetric if $(a,b)M$ and $a\wedge b=0$ imply $(b,a)M^*$.

Definition (1.13). Let L be a lattice with 0 and 1. A *complement* of an element $a\in L$ is an element $a'\in L$ such that $a\vee a'=1$ and $a\wedge a'=0$. L is called *complemented* when every element of L has a complement.

A lattice L is called *relatively complemented* when all the intervals $L[a,b]$, where $a<b$, are complemented.

Theorem (1.14). *A $\perp$-symmetric lattice L with 1 is M-symmetric if L satisfies the following condition:*

(1.14.1) *Every element a of L has a complement a' such that $(a,a')M$ and $(a',a)M^*$.*

Proof. Let a and b be elements of L. It follows from (1.14.1) that there exists a complement c of $a\wedge b$ such that $(a\wedge b,c)M$ and $(c,a\wedge b)M^*$. We shall prove that $(a,b)M$ is equivalent to $(a\wedge c,b\wedge c)M$. It follows from (1.3) that the intervals $L[a\wedge b,1]$ and $L[0,c]$ are isomorphic and that the elements a and b of $L[a\wedge b,1]$ correspond to the elements $a\wedge c$ and $b\wedge c$ of $L[0,c]$ respectively. If $(a,b)M$ holds in L, then it holds in $L[a\wedge b,1]$. Hence, by the above isomorphism, $(a\wedge c,b\wedge c)M$ holds in $L[0,c]$ and then it holds in L. Conversely, if $(a\wedge c,b\wedge c)M$ holds in L, then it holds in $L[0,c]$. By the above isomorphism, $(a,b)M$ holds in $L[a\wedge b,1]$. It follows from (1.4) that $(a,b)M$ holds in L.

In the same way we can show the equivalence of $(b,a)M$ and $(b\wedge c,a\wedge c)M$. Since L is $\perp$-symmetric and since $(a\wedge c)\wedge(b\wedge c)=(a\wedge b)\wedge c=0$, $(a\wedge c,b\wedge c)M$ implies $(b\wedge c,a\wedge c)M$. Consequently, $(a,b)M$ implies $(b,a)M$. □

Note that a complemented $\perp$-symmetric lattice is not necessarily M-symmetric (see Exercise 1.2).

Definition (1.15). By a *direct product* $L=\prod(L_\alpha;\alpha\in I)$ of a family of lattices L_α we mean the lattice $\{[a_\alpha;\alpha\in I];a_\alpha\in L_\alpha\}$ where the order $[a_\alpha;\alpha\in I]\leqq[b_\alpha;\alpha\in I]$ is defined by $a_\alpha\leqq b_\alpha$ for every $\alpha\in I$. Evidently, for two elements $a=[a_\alpha;\alpha\in I]$ and $b=[b_\alpha;\alpha\in I]$ of L we have

$$a\vee b=[a_\alpha\vee b_\alpha;\alpha\in I] \quad\text{and}\quad a\wedge b=[a_\alpha\wedge b_\alpha;\alpha\in I].$$

It is easy to show that L has 0 (resp. 1) if and only if every L_α has the least (resp. greatest) element.

Lemma (1.16). *Let $a=[a_\alpha]$ and $b=[b_\alpha]$ be elements of the direct product $L=\prod(L_\alpha;\alpha\in I)$ of lattices L_α. Then $(a,b)M$ in L if and only if $(a_\alpha,b_\alpha)M$ in L_α for every $\alpha\in I$.*

Proof. Assume $(a,b)M$ in L, and fix $\alpha \in I$. If $c_\alpha \leqq b_\alpha$ in L_α, then taking the element $c=[c_\beta; \beta\in I]$ where $c_\beta = b_\beta$ for $\beta \neq \alpha$, we have $c \leqq b$ in L. Hence by the assumption $(c\vee a)\wedge b = c\vee(a\wedge b)$ in L, whence $(c_\alpha \vee a_\alpha)\wedge b_\alpha = c_\alpha \vee (a_\alpha \wedge b_\alpha)$ in L_α. Therefore $(a_\alpha, b_\alpha)M$ in L_α. The converse statement can be easily proved. □

Lemma (1.17). *Let L be the direct product of lattices $L_\alpha(\alpha\in I)$. L is M-symmetric if and only if L_α is M-symmetric for every $\alpha\in I$. When L has 0, L is $\perp$-symmetric if and only if L_α is $\perp$-symmetric for every $\alpha \in I$.*

Proof. Assume that L is M-symmetric, and fix $\alpha\in I$. If $(a_\alpha, b_\alpha)M$ in L_α, then taking two elements $a=[a_\beta; \beta\in I]$ and $b=[b_\beta; \beta\in I]$ such that $a_\beta = b_\beta$ for $\beta\neq\alpha$, we have $(a,b)M$ in L by (1.16). Hence $(b,a)M$ by the assumption and hence $(b_\alpha, a_\alpha)M$ in L_α by (1.16). Therefore L_α is M-symmetric. The converse statement can be easily proved.

The last statement can be proved similarly, by putting $a_\alpha = b_\alpha = 0_\alpha$ for $\beta \neq \alpha$. □

Lemma (1.18). *Let L be the direct product of two lattices L_1 and L_2, and assume that L has 0. If L is weakly modular, then either both L_1 and L_2 are modular or one of L_1 and L_2 consists of only a zero element.*

Proof. If L_1 is not modular, then there exists a non-modular pair (a_1, b_1) in L_1. If L_2 had a non-zero element a_2, then putting $a=[a_1, a_2]$ and $b=[b_1, a_2]$ we would have $(a,b)M$, since L is weakly modular and since $a\wedge b = [a_1\wedge b_1, a_2]\neq 0$. Hence $(a_1, b_1)M$ by (1.16), a contradiction. Therefore L_2 consists of only a zero element. □

Exercise 1.1. In the following three lattices, find pairs which are not modular or not dual-modular. Show that the third lattice is M-symmetric and weakly modular.

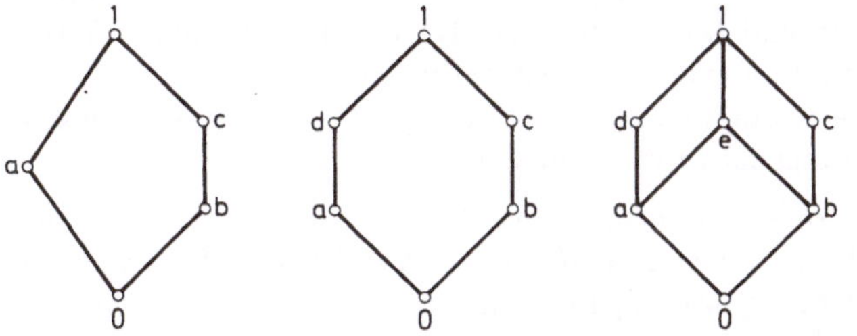

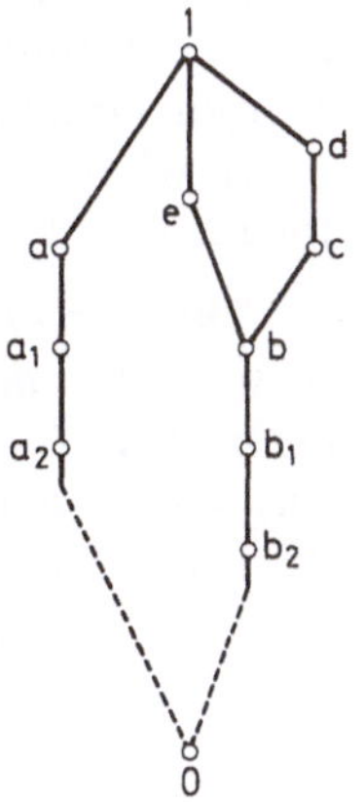

Exercise 1.2. Show that the fourth lattice is complemented and $\perp$-symmetric, but is not M-symmetric. ($L[0,a]$ and $L[0,b]$ are isomorphic to the lattice $\left\{0, \frac{1}{n}; n=1,2,\dots\right\}$ with the usual order.)

Exercise 1.3. Prove that if a lattice L with 0 and 1 satisfies (1.14.1) then so does every interval $L[a,b]$.

2. Semi-orthogonality in Lattices

As we shall see later, an orthogonality relation in a lattice satisfies the following four conditions: $a \perp a$ implies $a=0$; $a \perp b$ implies $b \perp a$; $a \perp b$, $a_1 \leqq a$ imply $a_1 \perp b$; $a \perp b$, $a \perp c$ imply $a \perp b \vee c$. But the fourth condition is so strong that we cannot give an orthogonality relation in a general symmetric lattice. In the following definition we take a weaker condition.

Definition (2.1). In a lattice L with 0, if there exists a binary relation "$\perp$" which satisfies the following axioms:

($\perp$1) $a \perp a$ implies $a=0$,
($\perp$2) $a \perp b$ implies $b \perp a$,
($\perp$3) $a \perp b$, $a_1 \leqq a$ imply $a_1 \perp b$,
($\perp$4) $a \perp b$, $a \vee b \perp c$ imply $a \perp b \vee c$,

then two elements a and b of L are said to be *semi-orthogonal* when $a \perp b$, and L is called a *semi-ortholattice.*

In a semi-ortholattice, $a \perp b$ implies $a \wedge b=0$; because, by ($\perp$2) and ($\perp$3), $a \perp b$ implies $a \wedge b \perp a \wedge b$ and hence $a \wedge b=0$ by ($\perp$1).

Definition (2.2). A family S of elements of a semi-ortholattice L is called a *semi-orthogonal family*, and write $(a; a \in S) \perp$, if for any pair of disjoint finite subsets F_1, F_2 of S we have

$$\bigvee(a; a \in F_1) \perp \bigvee(a; a \in F_2).$$

($\bigvee$ denotes the join of a collection of elements.)

Remark (2.3). Evidently any subset of a semi-orthogonal family forms a semi-orthogonal family. It is easily seen that a subset S is a semi-orthogonal family if every finite subset of S is.

In what follows, we denote by $A \cup B$ and $A \cap B$ the set-union and the set-intersection respectively of two sets A and B.

Theorem (2.4). *In a semi-ortholattice L, let S_i $(i \in I)$ be a collection of semi-orthogonal families of L, and assume that $\bigvee(a; a \in S_i)$ exists for every $i \in I$ (for instance, take S_i to be finite). If $(\bigvee(a; a \in S_i); i \in I) \perp$, then the set-union $\bigcup(S_i; i \in I)$ is also a semi-orthogonal family.*

Proof. (I) When $I=\{1,2\}$, let F_1 and F_2 be any disjoint finite subsets of $S_1 \cup S_2$, and put

$$b_{ij}=\bigvee(a; a \in F_i \cap S_j) \quad (i,j=1,2).$$

Since $b_{11} \perp b_{21}$ and $b_{11} \vee b_{21} \perp b_{22}$, it follows from ($\perp$4) that $b_{11} \perp b_{21} \vee b_{22}$. Similarly, since $b_{12} \perp b_{22}$ and $b_{12} \vee b_{22} \perp b_{11} \vee b_{21}$, we have $b_{12} \perp b_{11} \vee b_{21} \vee b_{22}$. Applying ($\perp$4) again we have $b_{21} \vee b_{22} \perp b_{11} \vee b_{12}$, that is $\bigvee(a; a\in F_1) \perp \bigvee(a; a\in F_2)$. Hence $S_1 \cup S_2$ is a semi-orthogonal family.

(II) When $I=\{1,\dots,n\}$, we can prove the theorem by mathematical induction. In fact, if $S_1\cup\cdots\cup S_{i-1}$ $(i\leqq n)$ is a semi-orthogonal family, then since

$$\bigvee(a; a\in S_1\cup\cdots\cup S_{i-1})=\bigvee(\bigvee(a; a\in S_j);\ 1\leqq j\leqq i-1)\perp\bigvee(a; a\in S_i),$$

it follows from (I) that $S_1\cup\cdots\cup S_{i-1}\cup S_i$ is a semi-orthogonal family.

(III) When I is an infinite set, let F be any finite subset of $\bigcup(S_i; i\in I)$. There exists a finite subset J of I such that $F\subset\bigcup(S_i; i\in J)$. Since it follows from (II) that $\bigcup(S_i; i\in J)$ is a semi-orthogonal family, so is F. Hence, by (2.3), $\bigcup(S_i; i\in I)$ is a semi-orthogonal family. □

Corollary (2.5). *Let $a_1, a_2, \dots$ be elements of a semi-ortholattice L. Then $(a_1,\dots,a_n)\perp$ holds if*

$$a_1\vee\cdots\vee a_{i-1}\perp a_i \quad \textit{for } i=2,\dots,n.$$

Also $(a_1,a_2,\dots)\perp$ holds if

$$a_1\vee\cdots\vee a_{i-1}\perp a_i \quad \textit{for every } i\geqq 2.$$

Proof. The first statement follows from (2.4) by mathematical induction. The second one follows from the first one and (2.3). □

Definition (2.6). A semi-ortholattice with 1 is called a *semi-orthocomplemented* lattice if for every element a there exists an element $a^\perp$ such that

$$1=a\vee a^\perp \quad \text{and } a\perp a^\perp.$$

$a^\perp$ is called a *semi-orthocomplement* of a.

A semi-ortholattice is called a *relatively semi-orthocomplemented* lattice if for every elements $a\leqq b$ there exists an element c such that

$$b=a\vee c \quad \text{and } a\perp c.$$

c is called a *relative semi-orthocomplement* of a in b.

Remark (2.7). A semi-orthocomplement of an element a is a complement of a. Note that it is not necessarily unique.

Lemma (2.8). *Let a and b be elements of a semi-orthocomplemented lattice L.*

(2.8.1) *If $a\perp b$ then there exists a semi-orthocomplement $b^\perp$ of b such that $a\leqq b^\perp$.*

(2.8.2) *If $a\leqq b$ and if $b^\perp$ is a semi-orthocomplement of b then there exists a semi-orthocomplement $a^\perp$ of a such that $b^\perp\leqq a^\perp$.*

Proof. (I) Let $a \perp b$ and let c be a semi-orthocomplement of $a \vee b$. Then we have $(a, b, c) \perp$ and $a \vee b \vee c = 1$. Hence $b^{\perp} = a \vee c$ is a semi-orthocomplement of b.

(II) Let $a \leqq b$ and let $b^{\perp}$ be a semi-orthocomplement of b. Since $b^{\perp} \perp a$, it follows from (I) that there exists a semi-orthocomplement $a^{\perp}$ of a such that $b^{\perp} \leqq a^{\perp}$. □

Theorem (2.9). *In a relatively semi-orthocomplemented lattice L, if* $a \perp b$ *then* $(a, b)M$.

Proof. Let $c \leqq b$. Since $(c \vee a) \wedge b \geqq c \vee (a \wedge b) = c$, there exists $d \in L$ such that

$$(c \vee a) \wedge b = c \vee d \quad \text{and} \quad c \perp d.$$

Since $c \vee d \leqq b \perp a$, it follows from $(\perp 3)$ and $(\perp 4)$ that $d \perp c \vee a \geqq d$, which implies $d = 0$ by $(\perp 1)$. Therefore $(a, b)M$ holds. □

Corollary (2.10). *A relatively semi-orthocomplemented lattice L is relatively complemented.*

Proof. If $a \leqq c \leqq b$, then there exists d such that $b = c \vee d$ and $c \perp d$. Since $(d, c)M$ we have $(a \vee d) \wedge c = a$. On the other hand $(a \vee d) \vee c = a \vee b = b$. Hence $a \vee d$ is a complement of c in $L[a, b]$. □

Lemma (2.11). *Let a and b be elements of a relatively semi-orthocomplemented lattice L with* 1.

(2.11.1) *If* $a \perp b$ *then there exists a semi-orthocomplement* $b^{\perp}$ *of b such that* $a = (a \vee b) \wedge b^{\perp}$.

(2.11.2) *If* $a \leqq b$ *and if c is a relatively semi-orthocomplement of a in b then there exists a semi-orthocomplement* $a^{\perp}$ *of a such that* $c = b \wedge a^{\perp}$.

Proof. (I) If $a \perp b$ then by (2.8.1) there exists $b^{\perp}$ such that $a \leqq b^{\perp}$. Since $(b, b^{\perp})M$ by (2.9), we have

$$(a \vee b) \wedge b^{\perp} = a \vee (b \wedge b^{\perp}) = a.$$

(II) If $a \vee c = b$ and $a \perp c$ then it follows from (I) that there exists $a^{\perp}$ such that $c = (c \vee a) \wedge a^{\perp} = b \wedge a^{\perp}$. □

Definition (2.12). A *complete lattice* is a lattice L in which for every subset S of L, the join $\bigvee(a; a \in S)$ and the meet $\bigwedge(a; a \in S)$ exist.

A lattice L is complete if every subset has its join in L; because for every subset S, the set of all lower bounds of S has its join, which is evidently the meet of S. Similarly, L is complete if every subset has its meet in L.

Theorem (2.13). *A relatively semi-orthocomplemented lattice L is complete if every semi-orthogonal family has its join in L.*

Proof. Let S be an arbitrary subset of L and let T be the set of all lower bounds of S. We denote by Φ the collection of all semi-orthogonal families R with $R \subset T$. Φ is a partially ordered set ordered by set-inclusion. If $\mathscr{C}$ is a chain in Φ, then the set-union $\bigcup(R; R \in \mathscr{C})$ belongs to Φ by (2.3). Hence by Zorn's lemma, Φ has a maximal family R^*. By the assumption the join $a = \bigvee(x; x \in R^*)$ exists in L, and evidently $a \in T$. If we had $x \nleqq a$ for some $x \in T$, then taking a relative semi-orthocomplement c of a in $a \vee x$, we would have $c > 0$, whence $c \notin R^*$. Moreover, $R^* \cup \{c\} \in \Phi$, since $c \leqq a \vee x \in T$ and $c \perp a$. This contradicts the maximality of R^*. Therefore we have $x \leqq a$ for every $x \in T$, which means that a is the meet of S. Hence L is complete. ▯

Definition (2.14). Let $\{a_\delta; \delta \in D\}$ be a family of elements of a complete lattice L, where D is a directed set. We write $a_\delta \uparrow a$ when $\delta_1 \leqq \delta_2$ implies $a_{\delta_1} \leqq a_{\delta_2}$ and $a = \bigvee(a_\delta; \delta \in D)$.

A complete lattice L is called an *upper continuous* (or $\wedge$-*continuous*) lattice if in L

(2.14.1) $\qquad a_\delta \uparrow a$ implies $a_\delta \wedge b \uparrow a \wedge b$ for every b.

Dually we can define $a_\delta \downarrow a$ and a *lower continuous* (or $\vee$-*continuous*) lattice. When a complete lattice L is both upper and lower continuous, L is called a *continuous* lattice.

Definition (2.15). A semi-orthogonality relation "$\perp$" in a complete lattice is said to be *ortho-continuous* when it satisfies the following condition:

(2.15.1) $\qquad$ If $a_\delta \uparrow a$ and if $a_\delta \perp b$ for every $\delta \in D$ then $a \perp b$.

Lemma (2.16). *Let S be a semi-orthogonal family in a complete semi-ortholattice L. If the semi-orthogonality relation is ortho-continuous, then for any pair of disjoint subsets S_1 and S_2 of S it follows that*

$$\bigvee(a; a \in S_1) \perp \bigvee(a; a \in S_2).$$

Proof. We may assume that S_1 and S_2 are infinite sets. If F_1 and F_2 are finite subsets of S_1 and S_2 respectively, then since F_1 and F_2 are disjoint, we have

$$\bigvee(a; a \in F_1) \perp \bigvee(a; a \in F_2).$$

Since the collection of all finite subsets F_1 of S_1 constitutes a directed system ordered by set-inclusion, we may write

$$\bigvee(a; a \in F_1) \uparrow \bigvee(a; a \in S_1)$$

and similarly $\bigvee(a; a \in F_2) \uparrow \bigvee(a; a \in S_2)$.

It follows from (2.15.1) that

$$\bigvee(a; a\in S_1) \perp \bigvee(a; a\in F_2) \quad \text{for every } F_2,$$

and applying (2.15.1) again we have

$$\bigvee(a; a\in S_1) \perp \bigvee(a; a\in S_2). \quad \square$$

EXERCISE 2.1. Prove that if L is a semi-orthocomplemented lattice such that every element a has a unique semi-orthocomplement $a^\perp$ then $a\to a^\perp$ is an involutive $(a^{\perp\perp}=a)$ dual-automorphism of L.

EXERCISE 2.2. Prove that if L is a relatively semi-orthocomplemented lattice satisfying the following condition:

$$a\wedge b=0 \quad \text{implies } a\perp b,$$

then L is a modular lattice.

3. Semi-orthogonality in ⊥-Symmetric Lattices

It is easy to show that a modular lattice with 0 is a semi-ortholattice when $a\perp b$ is defined by $a\wedge b=0$. This can be generalized as the following theorem.

Theorem (3.1). *A ⊥-symmetric lattice L is a semi-ortholattice, when we define the semi-orthogonality relation "$a\perp b$" by the following condition:*

(3.1.1) $$a\wedge b=0 \quad \textit{and } (a,b)M.$$

Proof. (⊥1) is evident. (⊥2) follows from (1.11.1). To prove (⊥3), let $a\perp b$ and $a_1\leqq a$. Then $a_1\wedge b\leqq a\wedge b=0$, and by (1.5.3) $(a_1,b)M$ holds. Thus $a_1\perp b$. To prove (⊥4), let $a\perp b$ and $a\vee b\perp c$, that is,

$$a\wedge b=0, \quad (a\vee b)\wedge c=0, \quad (b,a)M \quad \text{and } (c,a\vee b)M.$$

Then by (1.6), $(b\vee c,a)M$ holds and $(b\vee c)\wedge a=b\wedge a=0$. Hence we have $a\perp b\vee c$. $\square$

Remark (3.2). In the definition of a ⊥-symmetric lattice L, (1.11.1) means the symmetry of the relation "$a\perp b$" in (3.1).

Lemma (3.3). *Let S be a semi-orthogonal family in a ⊥-symmetric lattice L. For any pair of finite subsets F_1, F_2 of S,*

$$\bigvee(a; a\in F_1)\wedge\bigvee(a; a\in F_2)=\bigvee(a; a\in F_1\cap F_2).$$

Proof. Put $F_1' = F_1 - F_1 \cap F_2$. Since F_1' and F_2 are disjoint, we have $\bigvee(a; a \in F_1') \perp \bigvee(a; a \in F_2)$. Hence

$$\begin{aligned}\bigvee(a; a \in F_1 \cap F_2) &= \bigvee(a; a \in F_1 \cap F_2) \vee \{\bigvee(a; a \in F_1') \wedge \bigvee(a; a \in F_2)\} \\ &= \{\bigvee(a; a \in F_1 \cap F_2) \vee \bigvee(a; a \in F_1')\} \wedge \bigvee(a; a \in F_2) \\ &= \bigvee(a; a \in F_1) \wedge \bigvee(a; a \in F_2). \quad \square\end{aligned}$$

Remark (3.4). If S is a semi-orthogonal family of non-zero elements in a $\perp$-symmetric lattice, then it is easy to prove by (3.3) that there exists an isomorphism between the lattice $\mathscr{J}(S)$ of all finite subsets of S and the sublattice $\{\bigvee(a; a \in F); F \in \mathscr{J}(S)\}$ of L.

Lemma (3.5). *If a $\perp$-symmetric lattice L is upper continuous, then the semi-orthogonality relation $a \perp b$, defined by* (3.1.1), *is ortho-continuous.*

Proof. Let $a_\delta \uparrow a$ and let $a_\delta \perp b$ for every $\delta \in D$. By (2.14.1) we have $0 = a_\delta \wedge b \uparrow a \wedge b$, whence $a \wedge b = 0$. When $c \leqq b$, it follows from $(a_\delta, b)M$ that

$$c = (c \vee a_\delta) \wedge b \uparrow (c \vee a) \wedge b,$$

whence $(c \vee a) \wedge b = c = c \vee (a \wedge b)$. Hence $(a, b)M$ and thus $a \perp b$ holds. $\square$

Lemma (3.6). *If a $\perp$-symmetric lattice L with* 1 *satisfies the condition* (1.14.1), *then L is relatively semi-orthocomplemented.*

Proof. Let $a \leqq b$ in L. It follows from (1.14.1) that there exists a complement a' of a such that $a \perp a'$ and $(a', a)M^*$. Then $a \perp a' \wedge b$ and $a \vee (a' \wedge b) = (b \wedge a') \vee a = b \wedge (a' \vee a) = b$. Hence $a' \wedge b$ is a relative semi-orthocomplement of a in b. $\square$

Definition (3.7). Let a and b be elements in a lattice L with 0. An element b_1 is called a *left complement* within b of a in $a \vee b$ when

(3.7.1) $\quad a \vee b = a \vee b_1, \quad a \wedge b_1 = 0, \quad (b_1, a)M$ and $b_1 \leqq b$.

We call L a *left complemented* lattice when for every pair of elements a and b in L there exists such a left complement.

When $a \leqq b$, we omit the phrase "within b"; and when $a \vee b = 1$, we omit the phrase "in $a \vee b$".

Lemma (3.8). *In a lattice L with* 0, *if $(a, b)M$ and if b_1 is a left complement within b of a in $a \vee b$, then b_1 is a left complement of $a \wedge b$ in b.*

Proof. Let b_1 be a left complement within b of a in $a \vee b$. If $(a, b)M$, then we have

$$b_1 \vee (a \wedge b) = (b_1 \vee a) \wedge b = (a \vee b) \wedge b = b.$$

Since $(b_1, a)M$ and $b_1 \wedge a = 0$, $(b_1, a \wedge b)M$ holds by (1.5). Hence b_1 is a left complement of $a \wedge b$ in b. ▯

Theorem (3.9). *A left complemented lattice L is M-symmetric and is a relatively semi-orthocomplemented lattice.*

Proof. (I) Assume $(a,b)M$. Let b_1 be a left complement within b of a in $a \vee b$. Then by (3.8) we have $b = b_1 \vee (a \wedge b)$. Since $(a \wedge b, a)M$, $(b_1, a)M$ and $b_1 \wedge a \leqq a \wedge b$, it follows from (1.6) that $(b,a)M$ holds. Thus L is M-symmetric.

(II) Whe $a \leqq b$, (3.7.1) means that

$$b = a \vee b_1 \quad \text{and} \quad a \perp b_1 .$$

Hence L is a relatively semi-orthocomplemented lattice. ▯

Remark (3.10). We shall show that any complemented modular lattice L is left complemented. For $a, b \in L$, let c be a complement of $a \wedge b$ and put $b_1 = c \wedge b$. Then we have $b_1 \leqq b$ and $a \wedge b_1 = a \wedge b \wedge c = 0$. Moreover, since

$$(a \wedge b) \vee b_1 = (a \wedge b) \vee (c \wedge b) = \{(a \wedge b) \vee c\} \wedge b = 1 \wedge b = b,$$

we have

$$a \vee b_1 = a \vee (a \wedge b) \vee b_1 = a \vee b .$$

Hence b_1 is a left complement within b of a in $a \vee b$.

Theorem (3.11). *Let Λ be a given complemented modular lattice having the lattice operations $a \sqcup b$ and $a \sqcap b$. Let S be a fixed subset of $\Lambda - \{0, 1\}$ with the following two properties:*

(3.11.1) $$a \in S \quad \text{and} \quad 0 < b \leqq a \quad \text{imply} \quad b \in S,$$

(3.11.2) $$a, b \in S \quad \text{implies} \quad a \sqcup b \in S .$$

If in the set $L \equiv \Lambda - S$ we give the same order as Λ, then L is a weakly modular M-symmetric lattice where the lattice operations $a \vee b$ and $a \wedge b$ satisfy the following conditions:

(3.11.3) $$a \vee b = a \sqcup b \quad \text{for all } a, b \in L,$$

(3.11.4) $$a \wedge b = \begin{cases} a \sqcap b & \text{if } a \sqcap b \in L, \\ 0 & \text{if } a \sqcap b \in S . \end{cases}$$

Moreover, for $a, b \in L$

(3.11.5) $$(a,b)M \quad \text{in } L \quad \text{if and only if} \quad a \sqcap b \in L,$$

(3.11.6) $$a \perp b \quad \text{in } L \quad \text{if and only if} \quad a \sqcap b = 0 .$$

Proof. (I) To show that $a \vee b$ exists and (3.11.3) holds, it suffices to prove

(1) $$a \sqcup b \in L \quad \text{for all } a, b \in L.$$

If $a=0$, then $a \sqcup b = b \in L$. Suppose $a \neq 0$ and $a \sqcup b \in S$. Then since $0 < a \leqq a \sqcup b$, it follows from (3.11.1) that $a \in S$, contrary to $a \in L$. Thus (1) has been proved.

Next we shall show that for $a, b \in L$ if the element $a \wedge b$ of L is defined by (3.11.4) then it is just the meet of a and b in L. It is evident that $a \wedge b \leqq a, b$. Let $c \in L$ and $c \leqq a, b$. When $a \sqcap b \in L$, we have $c \leqq a \sqcap b = a \wedge b$. When $a \sqcap b \in S$, it follows from (3.11.1) that $c=0$. Hence $c \leqq a \wedge b$ always holds. Thus $a \wedge b$ is the meet of a and b in L. Therefore L is a lattice.

(II) Now we shall prove (3.11.5) which is equivalent to the following two statements:

(2) $$a \sqcap b \in L \quad \text{implies } (a, b) M \quad \text{in } L, \quad \text{and}$$

(3) $$a \sqcap b \in S \quad \text{implies } (a, b) \overline{M} \quad \text{in } L.$$

Let $a \sqcap b \in L$ and take $c \in L$ with $0 < c \leqq b$. By (3.11.4) and (3.11.3) we have

$$c \vee (a \wedge b) = c \vee (a \sqcap b) = c \sqcup (a \sqcap b).$$

On the other hand, since $0 < c \leqq (c \sqcup a) \sqcap b$ and $c \notin S$, it follows from (3.11.1) that $(c \sqcup a) \sqcap b \notin S$ and then

$$(c \vee a) \wedge b = (c \sqcup a) \wedge b = (c \sqcup a) \sqcap b.$$

Hence, by the modularity of Λ, $(a, b) M$ holds in L. Thus (2) has been proved.

Let $a \sqcap b \in S$. Then $a \wedge b = 0$, $a \neq 0$ and $b \neq 0$. Since Λ is relatively complemented by (3.10) and (3.11), there exists $c \in \Lambda$ such that

$$b = (a \sqcap b) \sqcup c \quad \text{and} \quad (a \sqcap b) \sqcap c = 0.$$

If $c \in S$ then by (3.11.2) we would have $b \in S$, contrary to $b \in L$. Hence $c \in L$, and by (3.11.3) we have

$$c \vee a = c \sqcup a = c \sqcup (a \sqcap b) \sqcup a = b \sqcup a = b \vee a,$$

whence $(c \vee a) \wedge b = b$. On the other hand, $c \vee (a \wedge b) = c$. But we have $b > c$ since $a \sqcap b \neq 0$. Hence $(a, b) M$ does not hold in L. Thus (3) has been proved.

(III) It follows from (3.11.4) and (3.11.5) that L is weakly modular and M-symmetric. It follows from (3.11.4) that $a \wedge b = 0$ if and only if $a \sqcap b \in S \cup \{0\}$. Hence (3.11.6) is implied from (3.11.5). □

Definition (3.12). When a weakly modular M-symmetric lattice L arises from a complemented modular lattice Λ in the manner described in (3.11), we call L a *Wilcox lattice* and call Λ the *modular extension* of L. This relation is denoted by $L \equiv \Lambda - S$.

An element in S is called an *imaginary element* for L, and when S has a greatest element i then it is called the *imaginary unit* for L.

Lemma (3.13). *A Wilcox lattice $L \equiv \Lambda - S$ is left complemented if and only if S satisfies the following condition:*

(3.13.1) *If $b \leqq a$ in Λ and if $a \notin S$ then there exists $c \notin S$ such that $a = b \sqcup c$ and $b \sqcap c = 0$.*

Proof. (I) Assume that L is left complemented. Let $b \leqq a$ in Λ and let $a \notin S$. If $b \in S$, then, since Λ is relatively complemented, there exists $c \in \Lambda$ such that $a = b \sqcup c$ and $b \sqcap c = 0$. Then we have $c \notin S$, because if $c \in S$ then by (3.11.2) we would have $a \in S$.

If $b \in L$, then since L is left complemented, there exists $c \in L$ such that $a = b \vee c$ and $b \perp c$. Then $c \notin S$ and it follows from (3.11.3) and (3.11.6) that $a = b \sqcup c$ and $b \sqcap c = 0$.

(II) Assume that S satisfies (3.13.1). Let $a, b \in L$. It follows from (3.13.1) that there exists $b_1 \in L$ such that

$$b = (a \sqcap b) \sqcup b_1 \quad \text{and} \quad (a \sqcap b) \sqcap b_1 = 0.$$

Then $b_1 \leqq b$ and

$$a \vee b_1 = a \sqcup b_1 = a \sqcup (a \sqcap b) \sqcup b_1 = a \sqcup b = a \vee b.$$

Moreover, since $a \sqcap b_1 = a \sqcap b \sqcap b_1 = 0$, it follows from (3.11.6) that $a \wedge b_1 = 0$ and $(b_1, a) M$. Therefore b_1 is a left complement within b of a in $a \vee b$. □

Remark (3.14). In the construction (3.11) of the Wolcox lattice $L \equiv \Lambda - S$, we may use (3.13.1) instead of (3.11.2). Then L is a weakly modular, left complemented lattice.

Lemma (3.15). *Let $L \equiv \Lambda - S$ be a Wilcox lattice. If Λ is complete, then so is L and then for any family $\{a_\alpha; \alpha \in I\}$ of elements of L*

(3.15.1) $$\bigvee (a_\alpha; \alpha \in I) = \bigsqcup (a_\alpha; \alpha \in I).$$

Proof. We shall show that if $a_\alpha \in L$ for every $\alpha \in I$ then the join $a = \bigsqcup (a_\alpha; \alpha \in I)$ belongs to L. This is evident when I is empty or when $a_\alpha = 0$ for every $\alpha \in I$. When $a_\alpha \neq 0$ for some $\alpha \in I$, it follows from (3.11.1) that a belongs to L. Hence a is the join $\bigvee (a_\alpha; \alpha \in I)$ in L, and hence L is complete. □

EXERCISE 3.1. Let L be a left complemented lattice with 1. Prove that if an element a of L has a unique semi-orthocomplement then a has a unique complement.

EXERCISE 3.2. Show that the third lattice of Exercise 1.1 is a Wilcox lattice.

EXERCISE 3.3. In a 3-dimensional space $R \times R \times R$ (R: the field of real numbers), let π be the plane $\{(\lambda, \mu, \nu) \in R \times R \times R; \nu = 1\}$. Show that all points and all lines on π together with the empty set and π form a left complemented Wilcox lattice whose modular extension is isomorphic to the lattice of all subspaces of $R \times R \times R$.

4. Distributivity and the Center of a Lattice

Definition (4.1). Let a, b and c be three elements of a lattice L. We write $(a,b,c)D$ if

(4.1.1) $$(a \vee b) \wedge c = (a \wedge c) \vee (b \wedge c).$$

We write $(a,b,c)D^*$ if

(4.1.2) $$(a \wedge b) \vee c = (a \vee c) \wedge (b \vee c).$$

When the distributive laws (4.1.1) and (4.1.2) hold for all permutations of a, b and c, then we say that $\{a,b,c\}$ is a *distributive triple* and write $(a,b,c)T$.

Obviously $(a,b,c)D$ is the same as $(b,a,c)D$ and $(a,b,c)D^*$ is the same as $(b,a,c)D^*$. Hence there are six different laws contained in the relation $(a,b,c)T$.

A lattice L is called *distributive* when $(a,b,c)D$ and $(a,b,c)D^*$ hold for all elements a, b and c of L.

Remark (4.2). It is easy to verify that in a modular lattice L if either $(a,b,c)D$ or $(a,b,c)D^*$ then $(a,b,c)T$.

Definition (4.3). Let a and b be elements of a lattice L with 0. We write $a \triangledown b$ if

(4.3.1) $$(x \vee a) \wedge b = x \wedge b \quad \text{for every } x \in L.$$

For a subset S of L we denote by $S^\triangledown$ the set of $a \in L$ such that $a \triangledown b$ for all $b \in S$.

Remark (4.4). If $a \triangledown b$ then $a \wedge b = 0$ and $(a,b)M$; because, putting $x = 0$ in (4.3.1) we get $a \wedge b = 0$, and if $c \leq b$ then we have

$$(c \vee a) \wedge b = c \wedge b = c = c \vee (a \wedge b).$$

Moreover, it is easy to show that $a \triangledown b$ is equivalent to the following statement:

(4.4.1) $$a \wedge b = 0 \quad \text{and} \quad (x, a, b)D \quad \text{for every} \quad x \in L.$$

Hence it follows from (4.2) that in a modular lattice with 0 the relation $a \triangledown b$ is symmetric.

Definition (4.5). A non-empty subset S of a lattice L is called an *ideal* of L when S satisfies the following two conditions:

(4.5.1) $$a \in S \quad \text{and} \quad b \leqq a \quad \text{imply} \quad b \in S,$$

(4.5.2) $$a \in S \quad \text{and} \quad b \in S \quad \text{imply} \quad a \vee b \in S.$$

Lemma (4.6). *In a lattice* L *with* 0,

(4.6.1) $$a \triangledown b, \quad a_1 \leqq a \quad \text{and} \quad b_1 \leqq b \quad \text{imply} \quad a_1 \triangledown b_1,$$

(4.6.2) $$a_1 \triangledown b \quad \text{and} \quad a_2 \triangledown b \quad \text{imply} \quad a_1 \vee a_2 \triangledown b, \quad \text{and}$$

(4.6.3) $$S^{\triangledown} \text{ is an ideal of } L \text{ for every subset } S \text{ of } L.$$

Proof. If $a \triangledown b$, $a_1 \leqq a$ and $b_1 \leqq b$, then for any $x \in L$ we have

$$(x \vee a_1) \wedge b_1 = (x \vee a_1) \wedge (x \vee a) \wedge b \wedge b_1 = (x \vee a_1) \wedge x \wedge b \wedge b_1 = x \wedge b_1.$$

Hence $a_1 \triangledown b_1$, and thus (4.6.1) has been proved. If $a_1 \triangledown b$ and $a_2 \triangledown b$, then for any $x \in L$ we have

$$(x \vee a_1 \vee a_2) \wedge b = (x \vee a_1) \wedge b = x \wedge b.$$

Hence $a_1 \vee a_2 \triangledown b$, and thus (4.6.2) has been proved. (4.6.3) follows from (4.6.1) and (4.6.2). □

Definition (4.7). In a lattice L with 0, let $S_1, \ldots, S_n$ be subsets of L each of which contains 0. We say that L is the *direct sum* of $S_1, \ldots, S_n$ and we write $L = S_1 \cup \cdots \cup S_n$ when

(4.7.1) every element $a \in L$ can be expressed in the form $a = a_1 \vee \cdots \vee a_n$, $a_i \in S_i$ $(i = 1, \ldots, n)$

and

(4.7.2) $$S_i \subset S_j^{\triangledown} \quad \text{for} \quad i \neq j.$$

Lemma (4.8). *If a lattice* L *with* 0 *is a direct sum of* $S_1, \ldots, S_n$, *then the expression* (4.7.1) *is unique and the sets* $S_1, \ldots, S_n$ *are ideals of* L.

Proof. (I) Let $a = a_1 \vee \cdots \vee a_n = b_1 \vee \cdots \vee b_n$, where $a_i \in S_i$ and $b_i \in S_i$ $(i = 1, \ldots, n)$. It follows from (4.6.3) and (4.7.2) that $b_2 \vee \cdots \vee b_n \in S_1^{\triangledown}$, whence $b_2 \vee \cdots \vee b_n \triangledown a_1$. Hence

$$a_1 = a \wedge a_1 = \{b_1 \vee (b_2 \vee \cdots \vee b_n)\} \wedge a_1 = b_1 \wedge a_1 \leqq b_1.$$

Similarly, we have $b_1 \leqq a_1$, and consequently $a_1 = b_1$. Generally we have $a_i = b_i$ for every i.

(II) We shall show that S_1 is an ideal of L. When $a \in S_1$ and $b \leqq a$, b is expressed in the form

$$b = b_1 \vee \cdots \vee b_n, \qquad b_i \in S_i.$$

If $i \neq 1$, then since $b_i \leqq b \leqq a$ and $a \in S_i^{\triangledown}$, we have

$$b_i = a \wedge b_i = 0.$$

Hence $b = b_1 \in S_1$. Next when $a, b \in S_1$, $a \vee b$ is expressed in the form

$$a \vee b = c_1 \vee \cdots \vee c_n, \qquad c_i \in S_i.$$

If $i \neq 1$, then since $a, b \in S_1 \subset S_i^{\triangledown}$, we have

$$c_i = (a \vee b) \wedge c_i = a \wedge c_i = 0.$$

Hence $a \vee b = c_1 \in S_1$. Therefore S_1 is an ideal. Generally S_i is an ideal for every i. □

Remark (4.9). It is easy to show that if L is the direct product of lattices $L_1, \ldots, L_n$ and if L has 0 then each L_i has a zero element 0_i and $L = S_1 \cup \cdots \cup S_n$ where $S_i = \{[0_1, \ldots, 0_{i-1}, a_i, 0_{i+1}, \ldots, 0_n]; a_i \in L_i\}$.

Conversely, it follows from (4.8) that if $L = S_1 \cup \cdots \cup S_n$ then L is isomorphic to the direct product $S_1 \ldots S_n$ by the mapping $a \to [a_1, \ldots, a_n]$ where $a = a_1 \vee \cdots \vee a_n$; because this mapping is obviously one-to-one and onto, and it is evident that the converse mapping preserves the order.

Definition (4.10). An element z of a lattice L is called a *neutral element* when $(z, a, b)T$ for all $a, b \in L$. Obviously $z \in L$ is neutral when $(z, a, b)D$, $(a, b, z)D$, $(z, a, b)D^*$ and $(a, b, z)D^*$ for all $a, b \in L$.

Lemma (4.11). *The set of all neutral elements of a lattice L is a distributive sublattice of L.*

Proof. Let z_1 and z_2 be neutral elements. We shall show that $z_1 \vee z_2$ is neutral. For any $a, b \in L$, $(z_1 \vee z_2, a, b)D$ holds since

$$\begin{aligned}(z_1 \vee z_2 \vee a) \wedge b &= (z_1 \wedge b) \vee \{(z_2 \vee a) \wedge b\} \\ &= (z_1 \wedge b) \vee (z_2 \wedge b) \vee (a \wedge b) \\ &= \{(z_1 \vee z_2) \wedge b\} \vee (a \wedge b).\end{aligned}$$

$(a, b, z_1 \vee z_2)D$ holds since

$$\begin{aligned}(a \vee b) \wedge (z_1 \vee z_2) &= \{(a \vee b) \wedge z_1\} \vee \{(a \vee b) \wedge z_2\} \\ &= (a \wedge z_1) \vee (b \wedge z_1) \vee (a \wedge z_2) \vee (b \wedge z_2) \\ &= \{a \wedge (z_1 \vee z_2)\} \vee \{b \wedge (z_1 \vee z_2)\}.\end{aligned}$$

$(z_1 \vee z_2, a, b)D^*$ holds since

$$\begin{aligned}\{(z_1 \vee z_2) \wedge a\} \vee b &= (z_1 \wedge a) \vee (z_2 \wedge a) \vee b \\ &= (z_1 \wedge a) \vee (z_1 \wedge b) \vee (z_2 \wedge a) \vee b \\ &= \{z_1 \wedge (a \vee b)\} \vee \{(z_2 \vee b) \wedge (a \vee b)\} \\ &= (z_1 \vee z_2 \vee b) \wedge (a \vee b).\end{aligned}$$

$(a, b, z_1 \vee z_2)D^*$ holds since

$$\begin{aligned}(a \wedge b) \vee z_1 \vee z_2 &= \{(a \vee z_1) \wedge (b \vee z_1)\} \vee z_2 \\ &= (a \vee z_1 \vee z_2) \wedge (b \vee z_1 \vee z_2).\end{aligned}$$

Hence $z_1 \vee z_2$ is neutral. Similarly $z_1 \wedge z_2$ is neutral. Therefore all neutral elements form a sublattice which is obviously distributive. □

Definition (4.12). An element z of a lattice L with 0 and 1 is called a *central element* when there exist two lattices L_1 and L_2 and an isomorphism between L and the direct product $L_1 L_2$ such that z corresponds to the element $[1_1, 0_2] \in L_1 L_2$. Evidently 0 and 1 are central elements.

The set of all central elements of L is called the *center* of L and is denoted by $Z(L)$. When $Z(L)$ is the set of 0 and 1 alone, L is called an *irreducible* lattice; otherwise L is called *reducible.*

Theorem (4.13). *Let z be an element of a lattice L with 0 and 1. The following five statements are equivalent.*

(α) *z is a central element.*

(β) *z is a neutral element having a complement.*

(γ) *There exists an element z' such that*

$$a = (a \wedge z) \vee (a \wedge z') = (a \vee z) \wedge (a \vee z') \text{ for every } a \in L.$$

(δ) *There exists an element z' such that $z \triangledown z'$, $z' \triangledown z$ and $a = (a \wedge z) \vee (a \wedge z')$ for every $a \in L$.*

(ε) *There exists an element z' such that $z \wedge z' = 0$, $(z, z')M$, $(z', z)M$ and $a = (a \wedge z) \vee (a \wedge z')$ for every $a \in L$.*

Proof. (α)⇒(β). If z is a central element of L then L is isomorphic to $L_1 L_2$ where z corresponds to the element $[1_1, 0_2]$. Since $[0_1, 1_2]$ is a complement of $[1_1, 0_2]$, z has a complement in L. It is easy to verify that $[1_1, 0_2]$ is a neutral element of $L_1 L_2$. For instance, we have $([1_1, 0_2], [a_1, a_2], [b_1, b_2])D$, since

$$([1_1, 0_2] \vee [a_1, a_2]) \wedge [b_1, b_2] = [1_1, a_2] \wedge [b_1, b_2] = [b_1, a_2 \wedge b_2]$$

and

$$\begin{aligned}&([1_1, 0_2] \wedge [b_1, b_2]) \vee ([a_1, a_2] \wedge [b_1, b_2]) \\ &= [b_1, 0_2] \vee [a_1 \wedge b_1, a_2 \wedge b_2] = [b_1, a_2 \wedge b_2].\end{aligned}$$

Therefore z is a neutral element of L.

$(\beta) \Rightarrow (\gamma)$. If a neutral element z has a complement z', then since $(z, z', a)D$ and $(z, z', a)D^*$ for every $a \in L$, we have $a = (a \wedge z) \vee (a \wedge z') = (a \vee z) \wedge (a \vee z')$ for every $a \in L$.

$(\gamma) \Rightarrow (\delta)$. For any $a \in L$, since $a = (a \vee z) \wedge (a \vee z')$, we have $a \wedge z' = (a \vee z) \wedge z'$ and $a \wedge z = (a \vee z') \wedge z$. Hence $z \triangledown z'$ and $z' \triangledown z$.

$(\delta) \Rightarrow (\alpha)$. Assume (δ), and put $L_1 = L[0, z]$ and $L_2 = L[0, z']$. Since $z \triangledown z'$, it follows from (4.6.1) that $L_1 \subset L_2^{\triangledown}$. Similarly $z' \triangledown z$ implies $L_2 \subset L_1^{\triangledown}$. Moreover, every element $a \in L$ can be expressed as

$$a = a_1 \vee a_2, \quad \text{where } a_1 = a \wedge z \in L_1 \quad \text{and } a_2 = a \wedge z' \in L_2.$$

Hence we have $L = L_1 \cup L_2$, and by (4.9) L is isomorphic to $L_1 L_2$ where z corresponds to $[1_1, 0_2]$. Therefore z is a central element.

$(\delta) \Rightarrow (\varepsilon)$. Since $a \triangledown b$ implies $a \wedge b = 0$ and $(a, b)M$ by (4.4), (δ) implies (ε).

$(\varepsilon) \Rightarrow (\delta)$. Assume (ε), and we shall show $z \triangledown z'$ and $z' \triangledown z$. For any $a \in L$, since $a = (a \wedge z) \vee (a \wedge z')$ we have $a \vee z = (a \wedge z') \vee z$. Since $(z, z')M$ and $z \wedge z' = 0$, we have

$$(a \vee z) \wedge z' = \{(a \wedge z') \vee z\} \wedge z' = (a \wedge z') \vee (z \wedge z') = a \wedge z'.$$

Therefore $z \triangledown z'$. Similarly we have $z' \triangledown z$. □

Remark (4.14). A central element z of a lattice L with 0 and 1 has a unique complement which is also a central element; because, if z_1' and z_2' are complements of z, then

$$z_1' = z_1' \wedge (z \vee z_2') = (z_1' \wedge z) \vee (z_1' \wedge z_2') = z_1' \wedge z_2' \leqq z_2',$$

and similarly $z_2' \leqq z_1'$. Evidently if z corresponds to $[1_1, 0_2]$ in $L_1 L_2$ then the complement of z corresponds to $[0_1, 1_2]$.

Theorem (4.15). *The center* $Z(L)$ *of a lattice* L *with* 0 *and* 1 *is a Boolean sublattice of* L. (A Boolean lattice means a complemented distributive lattice.)

Proof. Let z_1 and z_2 be central elements. It follows from (4.11) that $z_1 \vee z_2$ is neutral. Taking complements z_1' and z_2' of z_1 and z_2 respectively, it is easy to see that $z_1' \wedge z_2'$ is a complement of $z_1 \vee z_2$. Hence, by (4.13), $z_1 \vee z_2$ is central. Similarly $z_1 \wedge z_2$ is central. Therefore $Z(L)$ is a sublattice of L which is obviously a Boolean lattice. □

Remark (4.16). Let L be a lattice with 0 and 1. If $z_1, \ldots, z_n$ are central elements of L such that

$$z_1 \vee \cdots \vee z_n = 1 \quad \text{and } z_i \wedge z_j = 0 \quad \text{for } i \neq j,$$

then $L=L[0,z_1]\cup\cdots\cup L[0,z_n]$. This follows because, for any $a\in L$ we have

$$a=a\wedge(z_1\vee\cdots\vee z_n)=(a\wedge z_1)\vee\cdots\vee(a\wedge z_n),$$

and moreover $L[0,z_i]\subset L[0,z_j]^{\triangledown}$ since $z_i\wedge z_j=0$ implies $z_i\triangledown z_j$ by (4.4.1).

Conversely, if $L=S_1\cup\cdots\cup S_n$ then we shall show that there exist central elements $z_1,\ldots,z_n$ such that $S_i=L[0,z_i]$ for $i=1,\ldots,n$. By (4.7.1), there exist elements $z_1,\ldots,z_n$ such that $1=z_1\vee\cdots\vee z_n$ and $z_i\in S_i$. Since $z_2\vee\cdots\vee z_n\in S_1^{\triangledown}$ by (4.7.2), for any $a\in S_1$ we have

$$a=(z_1\vee z_2\vee\cdots\vee z_n)\wedge a=z_1\wedge a\leqq z_1.$$

Since S_1 is an ideal, we have $S_1=L[0,z_1]$. Generally $S_i=L[0,z_i]$. Moreover z_i is central by (4.9).

Therefore we may say that a finite direct sum decomposition of L corresponds to a finite decomposition of 1 in $Z(L)$.

Definition (4.17). Let L be a lattice with 0. L is called *semicomplemented* when for any element $a\in L$ (with $a\neq 1$ if 1 exists) there exists a non-zero element $b\in L$ such that $a\wedge b=0$. L is called a *section semicomplemented* lattice (for brevity, a *SSC* lattice) when the interval $L[0,a]$ is semicomplemented for every $a>0$, in other words, when L satisfies the following condition:

(4.17.1) If $a>b$ in L then there exists $c\in L$ such that $0<c\leqq a$ and $c\wedge b=0$.

A lattice L with 1 is called a *dual section semicomplemented* lattice (for brevity, a *SSC** lattice) when L^* is a SSC lattice, in other words, when L satisfies the following condition:

(4.17.2) If $a<b$ in L then there exists $c\in L$ such that $1>c\geqq a$ and $c\vee b=1$.

Evidently a relatively complemented lattice with 0 and 1 is SSC and SSC*.

Theorem (4.18). *Let a and b be elements of a SSC* lattice L with* 0. *The following three statements are equivalent.*

(α) $a\triangledown b$.
(β) $x\vee a=1$ *implies* $b\leqq x$.
(γ) $x=(x\vee a)\wedge(x\vee b)$ *for every* $x\in L$.

Proof. (α)$\Rightarrow$(β). If $x\vee a=1$, then by (α) we have

$$b=(x\vee a)\wedge b=x\wedge b\leqq x.$$

$(\beta) \Rightarrow (\gamma)$. For $x \in L$ we put $y = (x \vee a) \wedge (x \vee b)$. If $y > x$, then by (4.17.2) there exists $c \in L$ such that

$$1 > c \geqq x \quad \text{and} \quad c \vee y = 1.$$

Since $c \vee a = c \vee x \vee a \geqq c \vee y = 1$, we have $c \vee a = 1$. Similarly $c \vee b = 1$. By (β) $c \vee a = 1$ implies $b \leqq c$. Hence $c = c \vee b = 1$, a contradiction. Therefore $x = y = (x \vee a) \wedge (x \vee b)$.

$(\gamma) \Rightarrow (\alpha)$. For any $x \in L$ it follows from (γ) that

$$x \wedge b = (x \vee a) \wedge (x \vee b) \wedge b = (x \vee a) \wedge b.$$

Hence (α) holds. □

Corollary (4.19). *In a SSC* lattice L with* 0,

(4.19.1) *the relation* $a \triangledown b$ *is symmetric, and*

(4.19.2) *if* $a_\alpha \triangledown b$ *for every* $\alpha \in I$ *and if* $\bigvee(a_\alpha; \alpha \in I)$ *exists then* $\bigvee(a_\alpha; \alpha \in I) \triangledown b$.

Proof. (4.19.1) follows from the equivalence of (α) and (γ) in (4.18). It follows from the equivalence of (α) and (β) in (4.18) that $b \triangledown a_\alpha$ for every $\alpha \in I$ implies $b \triangledown \bigvee_\alpha a_\alpha$. Hence (4.19.2) holds by (4.19.1). □

Theorem (4.20). *An element z of a relatively complemented lattice L with* 0 *and* 1 *is a central element if and only if z has a unique complement.*

Proof. Assume that z has a unique complement z'. Let $x \vee z = 1$. Taking a complement y of $x \wedge z$ in $L[0, x]$, it is easy to show that y is a complement of z in L. Hence $z' = y \leqq x$. It follows from (4.18) that $x = (x \vee z) \wedge (x \vee z')$ for every $x \in L$. Since the dual L^* of L is relatively complemented and since z has a unique complement z' in L^*,we have $x = (x \vee z) \vee (x \vee z')$ in L^* as above. Hence in L, we have $x = (x \wedge z) \vee (x \wedge z')$ for every $x \in L$. By (4.13), z is a central element.

The converse follows from (4.14). □

EXERCISE 4.1. Prove (4.2).

EXERCISE 4.2. Let z be an element of a lattice L. Prove that the following three statements are equivalent.

- (α) z is a neutral element.
- (β) There exists two lattices L_1 and L_2 and there exists an isomorphism between L and a sublattice of $L_1 L_2$ such that z corresponds to $[1_1, 0_2]$.
- (γ) For any $a, b \in L$, the triple $\{z, a, b\}$ generates a distributive sublattice of L.

EXERCISE 4.3. Let a and b be elements of a relatively complemented lattice L with 0 (without 1). Prove that the following statements are equivalent.

(α) $a \triangledown b$.
(β) $b \leqq a \vee x$ implies $b \leqq x$.
(γ) $x = (x \vee a) \wedge (x \vee b)$ for every $x \in L$.

5. Centers of Complete Lattices

Lemma (5.1). *Let z be a central element of a complete lattice L. In L we have the following generalized distributive laws:*

$$\bigvee(a_\alpha;\alpha\in I)\wedge z = \bigvee(a_\alpha \wedge z;\alpha\in I), \tag{5.1.1}$$

$$\bigwedge(a_\alpha;\alpha\in I)\vee z = \bigwedge(a_\alpha \vee z;\alpha\in I). \tag{5.1.2}$$

Proof. It is evident that

$$\bigvee(a_\alpha;\alpha\in I)\wedge z \geqq \bigvee(a_\alpha \wedge z;\alpha\in I).$$

Let z' be the complement of z, and put

$$b = \bigvee(a_\alpha \wedge z;\alpha\in I)\vee z'$$

Since $a_\alpha = (a_\alpha \wedge z) \vee (a_\alpha \wedge z') \leqq b$ for every $\alpha \in I$, we have $\bigvee(a_\alpha;\alpha\in I) \leqq b$. Hence

$$\bigvee(a_\alpha;\alpha\in I)\wedge z \leqq b \wedge z = \bigvee(a_\alpha \wedge z;\alpha\in I).$$

Thus (5.1.1) is proved. (5.1.2) can be proved in a dual manner. □

Definition (5.2). In a complete lattice L, let $\{S_\alpha;\alpha\in I\}$ be an infinite family of subsets of L, where each of the subsets contains 0. We say that L is a *direct sum* of $\{S_\alpha;\alpha\in I\}$ and we write $L = \dot{\bigcup}(S_\alpha;\alpha\in I)$ when

(5.2.1) every element $a \in L$ can be expressed in the form

$$a = \bigvee(a_\alpha;\alpha\in I), \qquad a_\alpha \in S_\alpha \quad \text{for every } \alpha\in I,$$

and

$$S_\alpha \subset S_\beta^{\triangledown} \quad \text{for } \alpha \neq \beta. \tag{5.2.2}$$

Then we can show that each S_α is an ideal, in the same way as in the proof of (4.8). But, in order to prove the uniqueness of the expression (5.2.1), we need some assumption on L.

Remark (5.3). If $L = \dot{\bigcup} L[0, z_\alpha]$ where $z_\alpha \in Z(L)$, then the expression (5.2.1) is unique; because it follows from (5.1.1) that if $a = \bigvee(a_\alpha;\alpha\in I)$ where $a_\alpha \in L[0, z_\alpha]$, then

$$z_\alpha \wedge a = \bigvee(z_\alpha \wedge a_\beta;\beta\in I) = z_\alpha \wedge a_\alpha = a_\alpha.$$

Definition (5.4). A complete lattice is said to be $\triangledown$*-continuous* when the following statement holds in L:

(5.4.1) If $a_\delta \uparrow a$ and if $a_\delta \triangledown b$ for every $\delta \in D$ then $a \triangledown b$.

It follows from (4.6.2) that (5.4.1) is equivalent to

(5.4.2) If $a_\alpha \triangledown b$ for every $\alpha \in I$ then $\bigvee(a_\alpha; \alpha \in I) \triangledown b$.

Remark (5.5). It is easy to show that an upper continuous lattice is $\triangledown$-continuous. Moreover, it follows from (4.19.2) that a complete SSC* lattice is $\triangledown$-continuous.

Lemma (5.6). *Let* L *be a* $\triangledown$*-continuous lattice. If* $L = \dot{\bigcup}(S_\alpha; \alpha \in I)$, *then for every* $\alpha \in I$ *there exists* $z_\alpha \in Z(L)$ *such that* $S_\alpha = L[0, z_\alpha]$. *Moreover, every* $a \in L$ *is uniquely expressed as*

$$a = \bigvee(z_\alpha \wedge a; \alpha \in I), \qquad z_\alpha \wedge a \in S_\alpha \quad \textit{for every } \alpha.$$

Proof. First we shall prove the uniqueness of the expression (5.2.1). Let

$$a = \bigvee(a_\alpha; \alpha \in I) = \bigvee(b_\alpha; \alpha \in I) \quad \text{and } a_\alpha, b_\alpha \in S_\alpha \quad \text{for every } \alpha \in I.$$

Since if $\beta \neq \alpha$ $(\beta \in I)$ then $b_\beta \in S_\alpha^\triangledown$ by (5.2.2), we have $\bigvee(b_\beta; \beta \neq \alpha) \in S_\alpha^\triangledown$ by (5.4.2). Hence

$$a_\alpha = a \wedge a_\alpha = \{b_\alpha \vee \bigvee(b_\beta; \beta \neq \alpha)\} \wedge a_\alpha = b_\alpha \wedge a_\alpha \leqq b_\alpha.$$

Similarly, we have $b_\alpha \leqq a_\alpha$. Therefore the expression (5.2.1) is unique.

Next, by (5.2.1), there exist elements $\{z_\alpha; \alpha \in I\}$ such that $1 = \bigvee(z_\alpha; \alpha \in I)$ and $z_\alpha \in S_\alpha$. Since $\bigvee(z_\beta; \beta \neq \alpha) \in S_\alpha^\triangledown$ as above, for any $a \in S_\alpha$ we have

$$a = \{z_\alpha \vee \bigvee(z_\beta; \beta \neq \alpha)\} \wedge a = z_\alpha \wedge a \leqq z_\alpha.$$

Since S_α is an ideal, we have $S_\alpha = L[0, z_\alpha]$. Moreover, since the expression (5.2.1) is unique, L is isomorphic to the direct product $\prod(L[0, z_\alpha]; \alpha \in I)$ by the mapping $a \to [a_\alpha; \alpha \in I]$ where $a = \bigvee(a_\alpha; \alpha \in I)$. Putting $L_1 = L[0, z_\alpha]$ and $L_2 = \prod(L[0, z_\beta]; \beta \neq \alpha)$, we have an isomorphism between L and $L_1 L_2$ such that z_α corresponds to $[1_1, 0_2]$. Hence z_α is a central element.

The last statement of the lemma follows from (5.3). □

Definition (5.7). A complete lattice L is called a *Z-lattice* if L satisfies the following conditions:

(5.7.1) The center $Z(L)$ is a complete sublattice of L.

(5.7.2) If $z_\alpha \in Z(L)$ for every $\alpha \in I$ then $\bigvee(z_\alpha; \alpha \in I) \wedge a = \bigvee(z_\alpha \wedge a; \alpha \in I)$ for every $a \in L$.

Lemma (5.8). *Let L be a complete lattice satisfying* (5.7.2). *If z_α $(\alpha\in I)$ are central elements of L such that $\bigvee(z_\alpha;\alpha\in I)=1$ and $z_\alpha\wedge z_\beta=0$ for $\alpha\neq\beta$, then $L=\dot{\bigcup}(L[0,z_\alpha];\alpha\in I)$.*

Proof. It follows from (5.7.2) that for every $a\in L$ we have

$$a=\bigvee(z_\alpha;\alpha\in I)\wedge a=\bigvee(z_\alpha\wedge a;\alpha\in I),$$

where $z_\alpha\wedge a\in L[0,z_\alpha]$. Moreover, $z_\alpha\wedge z_\beta=0$ implies $z_\alpha\nabla z_\beta$ and hence implies $L[0,z_\alpha]\subset L[0,z_\beta]^\nabla$. Therefore L is the direct sum of $\{L[0,z_\alpha];\alpha\in I\}$. □

Remark (5.9). Let L be a ∇-continuous lattice satisfying (5.7.2). By (5.6) and (5.8), we may say that an infinite direct sum decomposition of L corresponds to an infinite decomposition of 1 in $Z(L)$.

Definition (5.10). Let a be an element of a Z-lattice L. It follows from (5.7.1) that there exists a unique least central element z such that $a\leqq z$. We call it the *central cover* of a and denote it by $e(a)$.

Lemma (5.11). *In a Z-lattice L,*

(5.11.1) $$e\left(\bigvee(a_\alpha;\alpha\in I)\right)=\bigvee(e(a_\alpha);\alpha\in I),\quad \textit{and}$$

(5.11.2) $$\textit{if } z\in Z(L) \textit{ then } e(z\wedge a)=z\wedge e(a).$$

Proof. (I) Put $a=\bigvee(a_\alpha;\alpha\in I)$. Since $e(a_\alpha)\leqq e(a)$ for every $\alpha\in I$, we have $\bigvee(e(a_\alpha);\alpha\in I)\leqq e(a)$. Since $a_\alpha\leqq e(a_\alpha)$, we have $a\leqq\bigvee(e(a_\alpha);\alpha\in I)$. By (5.7.1) $\bigvee(e(a_\alpha);\alpha\in I)$ is a central element, which implies

$$e(a)\leqq\bigvee(e(a_\alpha);\alpha\in I).$$

Thus (5.11.1) holds.

(II) Let $z\in Z(L)$ and z' be the complement of z. By (I)

$$e(a)=e(z\wedge a)\vee e(z'\wedge a).$$

Since $e(z\wedge a)\leqq z$ and $e(z'\wedge a)\leqq z'$, we have

$$z\wedge e(a)=\{z\wedge e(z\wedge a)\}\vee\{z\wedge e(z'\wedge a)\}=e(z\wedge a).\quad □$$

Remark (5.12). A continuous lattice L is a Z-lattice. Indeed, if $z_\delta\uparrow z$ and $z_\delta\in Z(L)$ then it is easy to show by the upper continuity that z is a neutral element. Taking the complement z'_δ of z_δ, we have $z'_\delta\downarrow z'$ for some element $z'\in L$. By the continuity, z' is a complement of z. Hence $z\in Z(L)$ by (4.13). If $z_\delta\downarrow z$ and $z_\delta\in Z(L)$ then we can prove $z\in Z(L)$ similarly. (5.7.2) follows from the upper continuity.

Theorem (5.13). *If a complete SSC lattice L is ∇-continuous, then L is a Z-lattice.*

Proof. Let $\{z_\alpha; \alpha\in I\}$ be a family of central elements, and for each $\alpha\in I$ let z'_α be the complement of z_α. Put $z=\bigvee(z_\alpha; \alpha\in I)$ and $z'=\bigwedge(z'_\alpha: \alpha\in I)$.

(I) Since $z_\alpha \triangledown z'_\alpha$, we have $z_\alpha \triangledown z'$ for every $\alpha\in I$. Hence by the $\triangledown$-continuity of L we have $z\triangledown z'$. Then $z\wedge z'=0$. We shall show $z'\triangledown z$. If we had $a\in L$ such that $(a\vee z')\wedge z>a\wedge z$, then by (4.17.1) there would exist $c\in L$ such that

$$0<c\leqq(a\vee z')\wedge z \quad \text{and } c\wedge a\wedge z=0.$$

Then $c\wedge a=c\wedge z\wedge a=0$. Since $c\leqq a\vee z'\leqq a\vee z'_\alpha$,

$$c=c\wedge(a\vee z'_\alpha)=(c\wedge a)\vee(c\wedge z'_\alpha)=c\wedge z'_\alpha\leqq z'_\alpha \quad \text{for every } \alpha\in I.$$

Hence $c\leqq z'$, whence $c\leqq z\wedge z'=0$, a contradiction. Thus $z'\triangledown z$ holds.

(II) Suppose $a>(a\wedge z)\vee(a\wedge z')$ for an element $a\in L$. By (4.17.1) there exists c such that

$$0<c\leqq a \quad \text{and } c\wedge\{(a\wedge z)\vee(a\wedge z')\}=0.$$

Then $c\wedge z=c\wedge a\wedge z=0$ and $c\wedge z'=c\wedge a\wedge z'=0$. Since $c\wedge z_\alpha=0$, we have $c=(c\wedge z_\alpha)\vee(c\wedge z'_\alpha)=c\wedge z'_\alpha\leqq z'_\alpha$ for every α, whence $c\leqq z'$. Hence $c=c\wedge z'=0$, a contradiction. Therefore

$$a=(a\wedge z)\vee(a\wedge z') \quad \text{for every } a\in L.$$

Together with (I), z satisfies the condition (δ) of (4.13). Thus $z\in Z(L)$.

(III) Put $\bar{z}=\bigwedge(z_\alpha; \alpha\in I)$ and $\bar{z}'=\bigvee(z'_\alpha; \alpha\in I)$. By the above arguments, $\bar{z}'$ is a central element and $\bar{z}$ is a complement of $\bar{z}'$. Hence $\bar{z}$ is a central element.

(IV) We shall prove that L satisfies (5.7.2), that is $z\wedge a=\bigvee_\alpha(z_\alpha\wedge a)$ for every $a\in L$. Suppose $z\wedge a>\bigvee_\alpha(z_\alpha\wedge a)$. By (4.17.1) there exists c such that

$$0<c\leqq z\wedge a \quad \text{and } c\wedge\bigvee_\alpha(z_\alpha\wedge a)=0.$$

Then since $c\wedge z_\alpha=c\wedge a\wedge z_\alpha=0$, we have

$$c=(c\wedge z_\alpha)\vee(c\wedge z'_\alpha)=c\wedge z'_\alpha\leqq z'_\alpha \quad \text{for every } \alpha,$$

whence $c\leqq z'$. Hence $c\leqq z\wedge z'=0$, a contradiction. □

Corollary (5.14). *If a complete lattice L is both SSC and SSC*, in particular if L is relatively complemented, then L is a Z-lattice.*

Proof. This follows from (5.5) and (5.13).

Exercise 5.1. Let L be a complete SSC lattice. Prove that if L satisfies (5.7.1) then it satisfies (5.7.2).

6. Perspectivity and Projectivity in Lattices

Definition (6.1). Let a and b be elements of a lattice L with 0. We say that a and b are *perspective* and we write $a \sim_x b$ or simply $a \sim b$ when

(6.1.1) $\quad a \vee x = b \vee x \quad$ and $a \wedge x = b \wedge x = 0 \quad$ for some $x \in L$.

It is evident that $a \sim_0 a$ for every a and that $a \sim 0$ implies $a = 0$.

We say that a is *subperspective* to b when

(6.1.2) $\quad a \leqq b \vee x \quad$ and $a \wedge x = 0 \quad$ for some $x \in L$.

It is evident that if $a \sim b_1 \leqq b$ then a is subperspective to b, and that if a is subperspective to b and if $a_1 \leqq a$ then a_1 is subperspective to b.

Lemma (6.2). *In a lattice with 0, if b is subperspective to a and if $a \triangledown b$ then $b = 0$. If $a \sim b$ and $a \triangledown b$ then $a = b = 0$.*

Proof. Let $b \leqq a \vee x$ and $b \wedge x = 0$. If $a \triangledown b$ then

$$b = (x \vee a) \wedge b = x \wedge b = 0.$$

The second statement follows from the first one. □

Lemma (6.3). *Let a and b be elements of a SSC lattice L. The following statements are equivalent.*

(α) $a \triangledown b$.
(β) *If $b_1 \leqq b$ and b_1 is subperspective to a, then $b_1 = 0$.*

Proof. (α)⇒(β). If $b_1 \leqq b$ and b_1 is subperspective to a, then since $a \triangledown b_1$ by (α) and (4.6.1), it follows from (6.2) that $b_1 = 0$.

(β)⇒(α). Assume that $a \triangledown b$ does not hold. There exists x such that $(x \vee a) \wedge b > x \wedge b$. By (4.17.1) there exists b_1 such that

$$0 < b_1 \leqq (x \vee a) \wedge b \quad \text{and} \quad b_1 \wedge x \wedge b = 0.$$

Then we have $b_1 \wedge x = 0$ since $b_1 \leqq b$. Hence b_1 is subperspective to a, which contradicts (β). □

Lemma (6.4). *Let a and b be elements of a relatively complemented lattice L with 0. If a is subperspective to b then there exists $b_1 \in L$ such that*

$$a \sim b_1 \leqq b.$$

Proof. Let $a \leqq b \vee x$ and $a \wedge x = 0$. Since $a \vee x \in L[x, b \vee x]$, there exists a complement y of $a \vee x$ in $L[x, b \vee x]$. Then $a \vee y = a \vee x \vee y = b \vee x$ and $a \wedge y = a \wedge (a \vee x) \wedge y = a \wedge x = 0$. Moreover, $b \vee y = b \vee x \vee y = b \vee x$. Taking a complement b_1 of $b \wedge y$ in $L[0, b]$, we have

$$b_1 \vee y = b_1 \vee (b \wedge y) \vee y = b \vee y = b \vee x \quad \text{and} \quad b_1 \wedge y = b_1 \wedge b \wedge y = 0.$$

Hence $a \sim_y b_1 \leqq b$. ▯

Theorem (6.5). *Let a and b be elements of a relatively complemented lattice L with* 0. *The following statements are equivalent.*

(α) $a \triangledown b$.
(β) *There do not exist non-zero elements* a_1 *and* b_1 *such that* $a_1 \sim b_1$, $a_1 \leqq a$ *and* $b_1 \leqq b$.

Proof. (α)⇒(β). If $a_1 \sim b_1$, $a_1 \leqq a$ and $b_1 \leqq b$, then since $a_1 \triangledown b_1$ by (α), we have $a_1 = b_1 = 0$ by (6.2).

(β)⇒(α). If $a \triangledown b$ does not hold, then by (6.3) there exists a non-zero element $b_1 \leqq b$ which is subperspective to a. By (6.4) there exists an element a_1 such that $b_1 \sim a_1 \leqq a$. This contradicts (β). ▯

Definition (6.6). Let a and b be elements of a lattice L with 0. We say that a and b are *projective* and we write $a \approx b$ when there exist elements $a_0, a_1, \ldots, a_n$ such that

$$a_0 = a, \quad a_n = b \quad \text{and } a_{i-1} \sim a_i \quad \text{for every } i = 1, \ldots, n.$$

Evidently, $a \approx b$ is an equivalence relation.

We say that a is *subprojective* to b when there exist elements $a_0, a_1, \ldots, a_n$ such that

$$a_0 = a, \quad a_n = b \quad \text{and } a_{i-1} \quad \text{is subperspective to } a_i \quad \text{for every } i.$$

Lemma (6.7). *In a lattice L with* 0,

(6.7.1) *if a is subprojective to b and if z is a neutral element, then* $b \leqq z$ *implies* $a \leqq z$.

In a Z-lattice,

(6.7.2) *if a is subprojective to b then* $e(a) \leqq e(b)$, *and*

(6.7.3) *if* $a \approx b$ *then* $e(a) = e(b)$.

Proof. (I) Let $a \leqq b \vee x$, $a \wedge x = 0$ and $b \leqq z$. Then

$$a = a \wedge (b \vee x) \leqq a \wedge (z \vee x) = (a \wedge z) \vee (a \wedge x) = a \wedge z \leqq z.$$

Applying this result successively, we get (6.7.1).

(II) If a subprojective to b, then since $b \leqq e(b)$, by (I) we have $a \leqq e(b)$, whence $e(a) \leqq e(b)$. Moreover, this implies (6.7.3). ▯

Lemma (6.8). *Let L be a complete lattice which is both SSC and SSC*. Then for each element* $a \in L$, *the join of all elements subprojective to a is equal to the central cover* $e(a)$.

Proof. The central cover $e(a)$ exists since L is a Z-lattice by (5.14). Let z be the join of all elements x subprojective to a. We have $z \leqq e(a)$, since $x \leqq e(a)$ by (6.7.2). Evidently $a \leqq z$. Hence, to prove the lemma, it suffices to show that z is a central element.

Put $z' = \bigvee(y \in L; z \triangledown y)$. By (4.19.1) and (4.19.2) we have $z' \triangledown z$ and $z \triangledown z'$. Assume that

$$b > (b \wedge z) \vee (b \wedge z') \quad \text{for some } b \in L.$$

By (4.17.1) there exists $c \in L$ such that

$$0 < c \leqq b \quad \text{and } c \wedge \{(b \wedge z) \vee (b \wedge z')\} = 0.$$

Then $c \wedge z = c \wedge b \wedge z = 0$ and $c \wedge z' = c \wedge b \wedge z' = 0$. Let x be an arbitrary element subprojective to a. If $c_1 \leqq c$ and c_1 is subperspective to x, then since c_1 is subprojective to a, we have $c_1 \leqq z$, whence $c_1 \leqq c \wedge z = 0$. Hence by (6.3) we have $x \triangledown c$, and hence $z \triangledown c$ by (4.19.2). By the definition of z' we have $c \leqq z'$, whence $c = c \wedge z' = 0$, a contradiction. Therefore

$$b = (b \wedge z) \vee (b \wedge z') \quad \text{for every } b \in L.$$

Thus, by (4.13) we have $z \in Z(L)$. □

Theorem (6.9). *Let L be a complete lattice which is both SSC and SSC*, and let a and b be elements of L. The following three statements are equivalent.*

(α) $e(a) \wedge e(b) = 0$.
(β) *If an element c is subprojective to both a and b, then* $c = 0$.
(γ) *If* $b_1 \leqq b$ *and if* b_1 *is subprojective to a, then* $b_1 = 0$.

Proof. (α)⇒(β). If $e(a) \wedge e(b) = 0$ and if c is subprojective to both a and b then by (6.7.2) we have $c \leqq e(a)$ and $c \leqq e(b)$, whence $c = 0$.

The implication (β)⇒(γ) is trivial.

(γ)⇒(α). Assume $e(a) \wedge e(b) \neq 0$. Since $e(e(a) \wedge b) = e(a) \wedge e(b) \neq 0$ by (5.11.2), we have $e(a) \wedge b \neq 0$. Then $x \triangledown b$ does not hold for some element x which is subprojective to a, since otherwise by (6.8) we would have $e(a) \triangledown b$, whence $e(a) \wedge b = 0$, a contradiction. By (6.3) there exists a non-zero element $b_1 \leqq b$ which is subperspective to x. Then b_1 is subprojective to a. This contradicts (γ). □

Corollary (6.10). *Let a and b be elements of a relatively complemented complete lattice L. The following statements are equivalent.*

(α) $e(a) \wedge e(b) = 0$.
(β) *There do not exist non-zero elements* a_1 *and* b_1 *such that* $a_1 \approx b_1$, $a_1 \leqq a$ *and* $b_1 \leqq b$.

Proof. If b_1 is subprojective to a, then applying (6.4) successively, we get an element a_1 such that $b_1 \approx a_1 \leqq a$. Hence (6.9) implies (6.10). □

EXERCISE 6.1. Let L be a relatively complemented lattice with 0 and 1. Prove that, in L, $a \sim b$ if and only if a and b have a common complement.

EXERCISE 6.2. Let L be a complete lattice which is both SSC and SSC*. Prove that an element z of L is central if and only if z contains every element which is subperspective to z.

PROBLEM 1. Is a $\perp$-symmetric lattice L M-symmetric if L is relatively semi-orthocomplemented?

References for Chapter I

For Section 1: G. Birkhoff [1] (Chapter IV, Sections 2 and 3), L. R. Wilcox [1], E. A. Schreiner [1], S. Maeda [7].

For Section 2: S. Maeda [1]; (2.13) is due to M. F. Janowitz [3].

For Section 3: L. R. Wilcox [1] and [2].

For Section 4: G. Birkhoff [1] (Chapter III), F. Maeda [2] (Chapter I), M. F. Janowitz [1] and [4].

For Section 5: F. Maeda [2] (Chapter I) and [3], M. F. Janowitz [4], S. Maeda [5].

For Section 6: M. F. Janowitz [4].

Chapter II

Atomistic Lattices and the Covering Property

In atomistic lattices, we shall consider an important property called the covering property. This property is weaker than $\perp$-symmetry but is very near to both $\perp$-symmetry and M-symmetry. In fact, if an atomistic lattice is either upper continuous or orthocomplemented then these three properties are equivalent (see (7.15) and (30.2)).

7. The Covering Property in Atomistic Lattices

Definition (7.1). In a lattice we say that b *covers* a and write $a \lessdot b$ when $a<b$ and moreover $a<c<b$ is not satisfied by any c.

An element p of a lattice with 0 is called an *atom* or a *point* when $0 \lessdot p$. An element h of a lattice with 1 is called a *dual-atom* or a *hyperplane* when $h \lessdot 1$.

A lattice L with 0 is called *atomic* when every non-zero element a of L contains an atom. L is called *atomistic* when every non-zero element a of L is the join of atoms contained in a.

The set of all atoms in an atomistic lattice L is called the *atom space* (or the *point space*) of L and we denote it by $\Omega(L)$.

Lemma (7.2). *A lattice L with 0 is atomistic if and only if L satisfies the following condition:*

(7.2.1) *If $a<b$ then there exists an atom p such that $p \nleqq a$ and $p \leqq b$* (i.e., L is *relatively atomic*).

Proof. Let L be atomistic. If $a<b$, then since $b=\bigvee(p;p\leqq b)$ there exists an atom $p\leqq b$ such that $p\nleqq a$. Hence L satisfies (7.2.1). Conversely, let L satisfy (7.2.1). For a non-zero element a, let S be the set of all atoms contained in a. If a were not the join of S, then there would exist an upper bound b of S such that $a\nleqq b$. Since $a\wedge b<a$, it follows from (7.2.1) that there exists an atom p such that $p\nleqq a\wedge b$ and $p\leqq a$. Since $p\in S$, we would have $p\leqq b$, whence $p\leqq a\wedge b$, a contradiction. Hence a is the join of S. $\square$

Lemma (7.3). *An atomic lattice L is atomistic if and only if L is section semicomplemented.*

Proof. An atomistic lattice is SSC, since (7.2.1) implies (4.17.1). Let L be an atomic SSC lattice, and let $a<b$ in L. By (4.17.1) there exists an element c such that $0<c\leqq b$ and $c\wedge a=0$. Since L is atomic there exists an atom p such that $p\leqq c$. Then $p\not\leqq a$ and $p\leqq b$. Therefore L is atomistic by (7.2). □

Definition (7.4). In a lattice with 0, the following property is called the *covering property:*

(7.4.1) If p is an atom and $a\wedge p=0$ then $a\lessdot a\vee p$.

Lemma (7.5). *Let a and b be elements of a lattice L.*

(7.5.1) *If $a\wedge b\lessdot b$ then $(a,b)M$.*

(7.5.2) *If $b\lessdot a\vee b$ then $(a,b)M^*$.*

(7.5.3) *If $a\lessdot a\vee b$ and $(a,b)M$ then $a\wedge b\lessdot b$.*

(7.5.4) *If $a\wedge b\lessdot a$ and $(a,b)M^*$ then $b\lessdot a\vee b$.*

Proof. By the duality, it suffices to prove (7.5.1) and (7.5.3).

Assume $a\wedge b\lessdot b$. If $a\wedge b\leqq c\leqq b$, then since either $c=a\wedge b$ or $c=b$, we have

$$(c\vee a)\wedge b=c\vee(a\wedge b).$$

Hence $(a,b)M$ in $L[a\wedge b, a\vee b]$. By (1.4), we have $(a,b)M$ in L.

Next, assume $a\lessdot a\vee b$ and $(a,b)M$. Since $a<a\vee b$, we have $a\wedge b<b$. If $a\wedge b\leqq c\leqq b$, they by $(a,b)M$ we have

$$(c\vee a)\wedge b=c\vee(a\wedge b)=c.$$

On the other hand, since $a\lessdot a\vee b$ and $a\leqq c\vee a\leqq a\vee b$, $c\vee a$ is equal to either a or $a\vee b$. Hence c is equal to either $a\wedge b$ or b. Therefore $a\wedge b\lessdot b$. □

Theorem (7.6). *Let L be a lattice with 0. The following three statements are equivalent.*

(α) *L has the covering property.*

(β) *If p is an atom of L then $(p,x)M$ for every $x\in L$.*

(γ) *If p is an atom of L then $(p,x)M^*$ for every $x\in L$.*

Proof. The equivalence of (β) and (γ) follows from (1.2).

(α)$\Rightarrow$(γ). Let p be an atom. If $x\wedge p\neq 0$, then since $p\leqq x$, we have $(p,x)M^*$. If $x\wedge p=0$, then by (α) we have $x\lessdot p\vee x$, whence $(p,x)M^*$ by (7.5.2).

(γ)$\Rightarrow$(α). Let $a\wedge p=0$. Since $p\wedge a=0\lessdot p$ and since $(p,a)M^*$ by (γ), we have $a\lessdot p\vee a$ by (7.5.4). □

Corollary (7.7). *Any $\perp$-symmetric lattice has the covering property.*

Proof. Let p be an atom of a $\perp$-symmetric lattice L. By (7.6) it suffices to prove $(p,a)M$ for every $a \in L$. If $p \leqq a$ then $(p,a)M$ holds evidently. If $p \nleqq a$, then $a \wedge p = 0$. Moreover, we have $(a,p)M$ by (7.5.1), since p is an atom. Therefore, we have $(p,a)M$, since L is $\perp$-symmetric. □

Definition (7.8). In a lattice with 0, the following property is called the *exchange property:*

(7.8.1) If p and q are atoms and if $a \wedge p = 0$ then $p \leqq a \vee q$ implies $q \leqq a \vee p$ (and hence implies $a \vee p = a \vee q$).

Lemma (7.9). *If a lattice L with 0 has the covering property then L has the exchange property.*

Proof. If $a \wedge p = 0$ and $p \leqq a \vee q$ where p and q are atoms, then $a \wedge q = 0$, since otherwise $q \leqq a$, whence $p \leqq a$, a contradiction. Hence $a \lessdot a \vee q$ by the covering property. Since $a < a \vee p \leqq a \vee q$, we have $a \vee p = a \vee q \geqq q$. □

Theorem (7.10). *Let L be an atomistic lattice. The following four statements are equivalent.*

(α) *L has the covering property.*
(β) *L has the exchange property.*
(γ) *$a \wedge b \lessdot a$ implies $b \lessdot a \vee b$ in L.*
(δ) *$a \wedge b \lessdot a$ implies $(a,b)M^*$ in L.*

Proof. (α) implies (β) by (7.9).

(β)⇒(γ). Let $a \wedge b \lessdot a$. It follows from $a \wedge b < a$ that $b < a \vee b$ and that there exists an atom p such that $p \nleqq a \wedge b$ and $p \leqq a$. Since $a \wedge b \lessdot a$, we have $(a \wedge b) \vee p = a$ and hence $b \vee p = a \vee b$. If $b < c \leqq a \vee b$, then there exists an atom q such that $q \nleqq b$ and $q \leqq c$. Since $q \leqq a \vee b = b \vee p$, by (β) we have $p \leqq b \vee q \leqq c$. Hence $c \geqq b \vee p = a \vee b$. Therefore $b \lessdot a \vee b$.

(γ) implies (δ) by (7.5.2), and (δ) implies (γ) by (7.5.4).

(γ)⇒(α). If p is an atom and if $a \wedge p = 0$ then $a \wedge p \lessdot p$. Hence by (γ) we have $a \lessdot a \vee p$. □

Definition (7.11). An atomistic complete lattice L is called a *compactly atomistic* lattice when L satisfies the following condition:

(7.11.1) If p is an atom and S is a set of atoms in L such that $p \leqq \bigvee(q; q \in S)$ then there exists a finite subset $\{q_1, \dots, q_n\}$ of S such that

$$p \leqq q_1 \vee \cdots \vee q_n.$$

Lemma (7.12). *Let p be an atom of a compactly atomistic lattice L. In L, if $p \leqq \bigvee(a_\alpha; \alpha \in I)$ then there exists a finite subset J of I such that $p \leqq \bigvee(a_\alpha; \alpha \in J)$.*

Proof. Put $S_\alpha = \{p \in \Omega(L); p \leqq a_\alpha\}$. Since L is atomistic, $a_\alpha = \bigvee(q; q \in S_\alpha)$. Hence

$$p \leqq \bigvee(a_\alpha; \alpha \in I) = \bigvee(q; q \in \bigcup(S_\alpha; \alpha \in I)).$$

It follows from (7.11.1) that there exists a finite subset F of $\bigcup(S_\alpha; \alpha \in I)$ such that $p \leqq \bigvee(q; q \in F)$. Taking a finite subset J of I such that $F \subset \bigcup(S_\alpha; \alpha \in J)$, we have

$$p \leqq \bigvee(q; q \in \bigcup(S_\alpha; \alpha \in J)) = \bigvee(a_\alpha; \alpha \in J). \quad \square$$

Lemma (7.13). *An atomistic complete lattice L is upper continuous if and only if L is compactly atomistic.*

Proof. (I) Assume that L is compactly atomistic. Let $a_\delta \uparrow a$ in L. For any $b \in L$ we have $a \wedge b = \bigvee(p \in \Omega(L); p \leqq a \wedge b)$. Hence, to prove $a_\delta \wedge b \uparrow a \wedge b$, it suffices to show that $p \leqq a \wedge b$ implies $p \leqq \bigvee(a_\delta \wedge b; \delta \in D)$. If $p \leqq a \wedge b$ then since $p \leqq a = \bigvee(a_\delta; \delta \in D)$, by (7.12) there exists a finite subset $\{\delta_1, \ldots, \delta_n\}$ such that $p \leqq a_{\delta_1} \vee \cdots \vee a_{\delta_n}$. Since D is a directed set, there exists $\delta_0 \in D$ such that $\delta_i \leqq \delta_0$ for every $i = 1, \ldots, n$. Then we have $p \leqq a_{\delta_0}$, and hence $p \leqq a_{\delta_0} \wedge b \leqq \bigvee(a_\delta \wedge b; \delta \in D)$. Thus L is upper continuous.

(II) Assume that L is upper continuous. In L, let p be an atom and S be a set of atoms such that $p \leqq \bigvee(q; q \in S)$. If we had $p \wedge \bigvee(q; q \in F) = 0$ for every finite subset F of S, then by the upper continuity we would have $p \wedge \bigvee(q; q \in S) = 0$, a contradiction. Hence there exists a finite subset $\{q_1, \ldots, q_n\}$ of S such that $p \wedge (q_1 \vee \cdots \vee q_n) \neq 0$, which implies $p \leqq q_1 \vee \cdots \vee q_n$. Hence L is compactly atomistic. $\square$

Lemma (7.14). *Let a and b be elements of an upper continuous lattice L. Then there exists a maximal element b_1 such that*

$$b_1 \leqq b, \quad b_1 \wedge a = 0 \quad \textit{and } (b_1, a) M.$$

Proof. Let S be the set of all $x \in L$ such that

$$x \leqq b, \quad x \wedge a = 0 \quad \text{and } (x, a) M.$$

S is a partially ordered set by the same order as L and $0 \in S$. To prove that S has a maximal element, by Zorn's lemma, it suffices to show that if C is a chain in S then $\bigvee(x; x \in C) \in S$. Put $c = \bigvee(x; x \in C)$. Obviously $c \leqq b$. Since L is upper continuous and since $\{x; x \in C\}$ is an increasing directed set, we have

$$c \wedge a = \bigvee(x \wedge a; x \in C) = 0.$$

If $a_1 \leqq a$, then since $(x,a)M$ and $x \wedge a = 0$, we have

$$(a_1 \vee x) \wedge a = a_1 \vee (x \wedge a) = a_1 \quad \text{for every } x \in C.$$

Hence, by the upper continuity of L, we have

$$(a_1 \vee c) \wedge a = a_1.$$

Thus $(c,a)M$ holds. Consequently we have $c \in S$. □

Theorem (7.15). *Let L be a compactly atomistic lattice. The following four statements are equivalent.*

(α) *L has the covering property.*
(β) *L is ⊥-symmetric.*
(γ) *L is M-symmetric.*
(δ) *L is left complemented.*

Proof. (δ) implies (γ) by (3.9). (γ) implies (β) evidently. (β) implies (α) by (7.7). We shall prove that (α) implies (δ). Take two elements a and b of L, where L has the covering property. It follows from (7.14) that there exists a maximal element b_1 such that

$$b_1 \leqq b, \quad b_1 \wedge a = 0 \quad \text{and} \quad (b_1, a)M.$$

To prove that b_1 is a left complement within b of a in $a \vee b$, it suffices to show $a \vee b_1 = a \vee b$. If we had $a \vee b_1 < a \vee b$, then since $a \vee b = a \vee \bigvee(p; p \leqq b)$, there would exist an atom p such that $p \leqq b$ and $p \not\leqq a \vee b_1$. Then $p \wedge (b_1 \vee a) = 0$ and it follows from (7.6) that $(p, b_1 \vee a)M$. Hence by (1.6)

$$(p \vee b_1, a)M \quad \text{and} \quad (p \vee b_1) \wedge a = b_1 \wedge a = 0.$$

Since $b_1 < p \vee b_1 \leqq b$, this contradicts the maximality of b_1. Thus we have $a \vee b_1 = a \vee b$. Consequently L is left complemented. □

Definition (7.16). A compactly atomistic M-symmetric lattice is called a *matroid lattice*. Since a matroid lattice is left complemented by (7.15), it is relatively complemented by (3.9) and (2.10).

Remark (7.17). "A complete atomic algebraic lattice" defined in Birkhoff [1] coincides with a compactly atomistic lattice in our sense. "An atomic matroid lattice" in Birkhoff [1] coincides with a matroid lattice in our sense.

Exercise 7.1. Prove that the covering property is equivalent to the following property:

If p is an atom and $p \not\leqq a \vee b$ then $a \wedge b = (a \vee p) \wedge b$.

EXERCISE 7.2. Show that the lattice of all topologies on a given set E is atomistic but has no covering property if E contains at least three elements. (For two topologies τ_1 and τ_2 on E, $\tau_1 \leqq \tau_2$ means that any τ_1-open set is τ_2-open.)

EXERCISE 7.3. Show that an atomistic lattice L which is not complete satisfies (7.11.1) if and only if $a_\delta \uparrow a$ implies $a_\delta \wedge b \uparrow a \wedge b$ in L (in other words, (7.13) can be proved without completeness of L).

8. Atomistic Lattices with the Covering Property

Definition (8.1). An element of a lattice L with 0 is called a *finite element* when it is either zero or the join of a finite number of atoms.

The set of all finite elements of L is denoted by $\mathscr{J}(L)$.

Lemma (8.2). *If a is a non-zero finite element, then there exists a finite family $\{p_1, \ldots, p_n\}$ of atoms such that*

(8.2.1) $a = p_1 \vee \cdots \vee p_n$ *and* $(p_1 \vee \cdots \vee p_{i-1}) \wedge p_i = 0$ *for* $i = 2, \ldots, n.$

Proof. There exist atoms $q_1, \ldots, q_m$ such that $a = q_1 \vee \cdots \vee q_m$. Starting from $i = 2$, we omit q_i in case $q_i \leqq q_1 \vee \cdots \vee q_{i-1}$. Thus we obtain a family of atoms with the desired property. □

Lemma (8.3). *Let p_i and q_i $(i = 1, \ldots, n)$ be atoms of a lattice L with the covering property. If*

$$(p_1 \vee \cdots \vee p_{i-1}) \wedge p_i = 0 \quad \text{for } i = 2, \ldots, n \quad \text{and if}$$
$$p_i \leqq q_1 \vee \cdots \vee q_n \quad \text{for } i = 1, \ldots, n,$$

then $p_1 \vee \cdots \vee p_n = q_1 \vee \cdots \vee q_n$.

Proof. We shall prove the lemma by mathematical induction. When $n = 1$, the lemma is trivial. Assuming that the lemma holds when $n = m - 1$, we shall show that it holds when $n = m$.

Let $(p_1 \vee \cdots \vee p_{i-1}) \wedge p_i = 0$ for $i = 2, \ldots, m$ and let

(1) $$p_i \leqq q_1 \vee \cdots \vee q_m \quad \text{for } i = 1, \ldots, m.$$

If $p_i \leqq q_1 \vee \cdots \vee q_{m-1}$ for every $i = 1, \ldots, m$, then by the assumption for $n = m - 1$ we would have

$$p_1 \vee \cdots \vee p_{m-1} = q_1 \vee \cdots \vee q_{m-1} \geqq p_m,$$

which contradicts $(p_1 \vee \cdots \vee p_{m-1}) \wedge p_m = 0$. Hence there exists $p_{i(1)}$ such that $p_{i(1)} \not\leqq q_1 \vee \cdots \vee q_{m-1}$. Then since

$$q_1 \vee \cdots \vee q_{m-1} < q_1 \vee \cdots \vee q_{m-1} \vee p_{i(1)} \leqq q_1 \vee \cdots \vee q_m,$$

by the covering property we have $q_1 \vee \cdots \vee q_{m-1} \vee p_{i(1)} = q_1 \vee \cdots \vee q_m$. Hence

(2) $$p_i \leqq q_1 \vee \cdots \vee q_{m-1} \vee p_{i(1)} \quad \text{for } i=1, \ldots, m.$$

Starting from (2) instead of (1), there exists $p_{i(2)}$ such that $p_{i(2)} \nleqq q_1 \vee \cdots \vee q_{m-2} \vee p_{i(1)}$. Then we have $i(1) \neq i(2)$ and

$$q_1 \vee \cdots \vee q_{m-2} \vee p_{i(1)} \vee p_{i(2)} = q_1 \vee \cdots \vee q_{m-1} \vee p_{i(1)} = q_1 \vee \cdots \vee q_m.$$

Continuing this process, finally we obtain $p_{i(m)}$ such that $p_{i(m)} \nleqq p_{i(1)} \vee \cdots \vee p_{i(m-1)}$. Then we have $i(m) \neq i(1), \ldots, i(m-1)$ and

$$p_{i(1)} \vee \cdots \vee p_{i(m)} = q_1 \vee p_{i(1)} \vee \cdots \vee p_{i(m-1)} = \cdots = q_1 \vee \cdots \vee q_m.$$

Therefore $p_1 \vee \cdots \vee p_m = q_1 \vee \cdots \vee q_m$. □

Theorem (8.4). *Let a be a non-zero finite element of a lattice L with the covering property. The number of atoms $\{p_i\}$ satisfying* (8.2.1) *is uniquely determined.*

Proof. Let $a = p_1 \vee \cdots \vee p_m = q_1 \vee \cdots \vee q_n$, $(p_1 \vee \cdots \vee p_{i-1}) \wedge p_i = 0$ $(i=2, \ldots, m)$ and $(q_1 \vee \cdots \vee q_{j-1}) \wedge q_j = 0$ $(j=2, \ldots, n)$. If $m > n$, then by (8.3)

$$a > p_1 \vee \cdots \vee p_n = q_1 \vee \cdots \vee q_n = a,$$

a contradiction. Hence we have $m \leqq n$. By symmetry, $m = n$. □

Definition (8.5). Let L be a lattice with the covering property. If a is a non-zero finite element of L, the number determined by (8.4) is called the *height* or the *dimension* of a and is denoted by $h(a)$. We define $h(0)=0$, and define $h(a)=\infty$ if $a \notin \mathscr{I}(L)$.

Evidently, $p \in L$ is an atom if and only if $h(p)=1$.

Remark (8.6). Let L be a lattice with the covering property. For $a \in \mathscr{I}(L)$, if $h(a)=n$ and (8.2.1) holds then putting $a_i = p_1 \vee \cdots \vee p_i$, we have the following connected chain with length n.

(8.6.1) $$0 \lessdot a_1 \lessdot a_2 \lessdot \cdots \lessdot a_n = a.$$

Conversely (8.6.1) implies $h(a)=n$, provided that L is atomistic. For, there exist atoms $p_1, \ldots, p_n$ such that $p_i \wedge a_{i-1} = 0$ and $p_i \vee a_{i-1} = a_i$, and then $p_1, \ldots, p_n$ satisfy (8.2.1).

Definition (8.7). For brevity, we call an atomistic lattice with the covering property an *AC-lattice*. A matroid lattice may be called an upper continuous AC-lattice.

Let L be an AC-lattice with 1. The height $h(1)$ is called the *length* of L.

It is easy to show that if an AC-lattice L has no unit element then the set $\{h(a); a\in\mathcal{J}(L)\}$ is unbounded, even if $L=\mathcal{J}(L)$.

Lemma (8.8). *If L is an AC-lattice, then $\mathcal{J}(L)$ is an ideal of L, and for $a,b\in\mathcal{J}(L)$, $a<b$ implies $h(a)<h(b)$.*

Proof. It is evident that $a,b\in\mathcal{J}(L)$ implies $a\vee b\in\mathcal{J}(L)$. Let $a<b$ and $b\in\mathcal{J}(L)$. There exist atoms $q_1,\dots,q_n$ such that $b=q_1\vee\cdots\vee q_n$ and $h(b)=n$. If we had $n\leqq h(a)$ ($\leqq\infty$), then there would exist atoms $p_1,\dots,p_n$ such that $p_i\leqq a$ $(i=1,\dots,n)$ and $(p_1\vee\cdots\vee p_{i-1})\wedge p_i=0$ $(i=2,\dots,n)$. By (8.3)

$$a\geqq p_1\vee\cdots\vee p_n=q_1\vee\cdots\vee q_n=b,$$

a contradiction. Hence $a\in\mathcal{J}(L)$ and $h(a)<n$. □

Remark (8.9). Let L be an AC-lattice. It follows from (8.6) that the lattice $\mathcal{J}(L)$ satisfies the Jordan-Dedekind chain condition. That is to say, when $a<b$ in $\mathcal{J}(L)$, all connected chains between a and b have the same length $h(b)-h(a)$.

Lemma (8.10). *If an AC-lattice L with 1 is of finite length then L is compactly atomistic.* (Thus L is a matroid lattice of finite length.)

Proof. (I) To prove that L is complete, first we shall show that if S is a set of atoms of L then there exists a finite subset F of S such that $\bigvee(p;p\in F)$ is a least upper bound of S. Since $h(\bigvee(p;p\in F))\leqq h(1)$ for every finite subset F of S, we may take F such that $h(\bigvee(p;p\in F))$ is the largest. Then, for any $q\in S$ we have $q\leqq\bigvee(p;p\in F)$, for otherwise $h(q\vee\bigvee(p;p\in F))>h(\bigvee(p;p\in F))$, a contradiction. Hence $\bigvee(p;p\in F)$ is a least upper bound of S. This means that $\bigvee(p;p\in S)$ exists and is equal to $\bigvee(p;p\in F)$.

(II) For any subset $\{a_\alpha;\alpha\in I\}$ of L, it follows from (I) that the join $\bigvee(p\in\Omega(L);\ p\leqq a_\alpha$ for some $\alpha\in I)$ exists. Since L is atomistic, this is just the join of $\{a_\alpha\}$. Therefore L is complete. Moreover, it immediately follows from (I) that L satisfies (7.11.1). Hence L is compactly atomistic. □

Theorem (8.11). *If L is an AC-lattice then the ideal $\mathcal{J}(L)$ of all finite elements of L is a left complemented lattice and it is M-symmetric.*

Proof. By the definition of a left complemented lattice, it suffices to show that the interval $L[0,a]$ is left complemented for every $a\in\mathcal{J}(L)$. Since $L[0,a]$ is an AC-lattice of finite length, it follows from (8.10) and (7.15) that $L[0,a]$ is left complemented. Hence $\mathcal{J}(L)$ is left complemented and by (3.9) it is M-symmetric. □

Remark (8.12). Let L be an AC-lattice. By (8.11) and (3.1), $\mathscr{F}(L)$ has a semi-orthogonal relation $a \perp b$ defined by the conditions: $a \wedge b = 0$ and $(a,b)M$. If p is an atom, then evidently $a \wedge p = 0$ implies $a \perp p$. It follows from (2.5) that $h(a) = n$ if and only if there exists a semi-orthogonal family of atoms $p_1, \ldots, p_n$ such that $a = p_1 \vee \cdots \vee p_n$. Moreover $\mathscr{F}(L)$ is relatively semi-orthocomplemented by (3.9).

Lemma (8.13). *Let L be an AC-lattice. Then*

$$h(a \vee b) \leqq h(a) + h(b) \quad \text{for every } a, b \in \mathscr{F}(L),$$

and equality holds if $a \perp b$.

Proof. If $a \perp b$ $(a, b \in \mathscr{F}(L))$ then by the property (2.4) of semi-orthogonal families we have $h(a \vee b) = h(a) + h(b)$. For any $a, b \in \mathscr{F}(L)$, since $\mathscr{F}(L)$ if left complemented there exists b_1 such that

$$b_1 \leqq b, \qquad a \vee b_1 = a \vee b \quad \text{and } a \perp b_1.$$

Hence $h(a \vee b) = h(a) + h(b_1) \leqq h(a) + h(b)$. □

Theorem (8.14). *Let L be an AC-lattice. Then*

$$h(a \vee b) + h(a \wedge b) \leqq h(a) + h(b) \quad \text{for every } a, b \in \mathscr{F}(L),$$

and equality holds if and only if $(a, b)M$.

Proof. (I) Let $a, b \in \mathscr{F}(L)$. Since $\mathscr{F}(L)$ is left complemented, there exists a_1 such that

$$a = (a \wedge b) \vee a_1 \quad \text{and } a \wedge b \perp a_1.$$

Since $a_1 \vee b = a_1 \vee (a \wedge b) \vee b = a \vee b$, by (8.13) we have

$$h(a \vee b) \leqq h(a_1) + h(b) \quad \text{and } h(a) = h(a \wedge b) + h(a_1).$$

Hence $h(a \vee b) + h(a \wedge b) \leqq h(a) + h(b)$.

(II) Assume $(a,b)M$. If b_1 is a left complement within b of a in $a \vee b$ then by (3.8) b_1 is a left complement of $a \wedge b$ in b. Hence

$$h(a \vee b) + h(a \wedge b) = h(a) + h(b_1) + h(a \wedge b) = h(a) + h(b).$$

(III) Assume $(a, b)\bar{M}$. Then there exists $c \leqq b$ such that

$$(c \vee a) \wedge b > c \vee (a \wedge b).$$

Put $x = (c \vee a) \wedge b$ and $y = c \vee (a \wedge b)$. Since $a \wedge b \leqq a \wedge y \leqq a \wedge x = a \wedge b$, we have $a \wedge y = a \wedge b$. By (I) we have

$$h(a) + h(y) \geqq h(a \vee y) + h(a \wedge y) = h(c \vee a) + h(a \wedge b) \quad \text{and}$$
$$h(b) + h(c \vee a) \geqq h(a \vee b) + h(x).$$

Hence $h(a)+h(b) \geqq h(c \vee a)+h(a \wedge b)-h(y)+h(a \vee b)+h(x)-h(c \vee a)$
$$=h(a \vee b)+h(a \wedge b)+h(x)-h(y)>h(a \vee b)+h(a \wedge b). \quad \square$$

Definition (8.15). For a non-zero element a of an AC-lattice L, if P is a semi-orthogonal family of atoms such that $a=\bigvee(p; p \in P)$, then we say that P is a *base* of a.

If a is finite then it follows from (8.2) and (8.4) that a has at least one base and all bases of a have the same cardinal number equal to the height $h(a)$.

Lemma (8.16). *Let a be a non-zero element of a complete AC-lattice L, and let Q be a semi-orthogonal family of atoms such that $p \leqq a$ for every $p \in Q$. Then there exists a base P of a such that $Q \subset P$. In particular, any non-zero element of L has a base.*

Proof. Let S be the set of all atoms contained in a. We have $Q \subset S$ and $a=\bigvee(p; p \in S)$. Denote by Φ the set of all semi-orthogonal families R of atoms such that $Q \subset R \subset S$. Φ is a partially ordered set ordered by set-inclusion. If $\mathscr{C}$ is a chain in Φ, then the set union $\bigcup(R; R \in \mathscr{C})$ belongs to Φ since it is a semi-orthogonal family by (2.3). Hence by Zorn's lemma, Φ has a maximal element P. If we had $\bigvee(p; p \in P)<a$, then there would exist an atom $q \in S$ such that $q \not\leqq \bigvee(p; p \in P)$. Since $\bigvee(p; p \in F) \perp q$ for every finite subset F of P, by (2.3) we would have $P \cup \{q\} \in \Phi$, which contradicts the maximality of P. Hence P is the required base of a. $\square$

Remark (8.17). In a lattice L with 0, the following property may be called the *finite covering property:*

(8.17.1) If $p \in \Omega(L)$, $a \in \mathscr{I}(L)$ and $a \wedge p=0$ then $a \lessdot a \vee p$.

Though this property is weaker than the covering property, it is easy to prove that a compactly atomistic lattice with this property is a matroid lattice.

It is evident that the results from (8.3) to (8.6) are valid when L has the finite covering property. Hence the results from (8.8) to (8.16) are valid when L is an atomistic lattice with the finite covering property.

Lemma (8.18). *Let $a<b$ in an AC-lattice L. Then $L[a,b]$ is an AC-lattice and an element $c \in L$ is an atom of $L[a,b]$ if and only if there exists an atom p of L such that*

(8.18.1) $$c=a \vee p, \quad a \wedge p=0 \quad \textit{and} \quad p \leqq b.$$

The same statement holds for the sublattice $L[a, \rightarrow]$.

Proof. (I) By the covering property, (8.18.1) implies
$$a \lessdot a \vee p=c \leqq b.$$

Hence c is an atom of $L[a,b]$. Conversely, if c is an atom of $L[a,b]$, then taking an atom p of L such that $p \nleqq a$ and $p \leqq c$, we have $a \wedge p = 0$ and $p \leqq b$. Moreover $a \lessdot c$ implies $c = a \vee p$.

(II) If $x < y$ in $L[a,b]$, then there exists an atom p of L such that $p \nleqq x$ and $p \leqq y$. Since $p \nleqq a$, $c = a \vee p$ is an atom of $L[a,b]$ by (I). Evidently $c \nleqq x$ and $c \leqq y$. Hence $L[a,b]$ is atomistic by (7.2). Moreover since the statement (γ) in (7.10) holds in L, it holds also in $L[a,b]$, whence $L[a,b]$ has the covering property. □

Lemma (8.19). *An AC-lattice L with* 1 *is dual-atomistic* (that is, L^* is atomistic) *if and only if L is dual section semicomplemented.*

Proof. If L is dual-atomistic, then by (7.3) L^* is SSC, whence L is SSC*. Conversely, assume that L is SSC*, and let $a < b$ in L^*. Then since $a > b$ in L, there exists an atom p of L such that $p \leqq a$ and $p \nleqq b$. Since $b < b \vee p$, by (4.17.2) there exists $h \in L$ such that

$$1 > h \geqq b \quad \text{and} \quad h \vee (b \vee p) = 1.$$

Then $h \vee p = h \vee b \vee p = 1$. We have $p \nleqq h$, since otherwise $h = h \vee p = 1$, a contradiction. By the covering property we have $h \lessdot h \vee p = 1$, whence h is a dual-atom in L. We have $h \ngeqq a$, since $p \leqq a$ and $p \nleqq h$. Therefore, in L^*, h is an atom such that $h \leqq b$ and $h \nleqq a$. Thus L^* is atomistic. □

Remark (8.20). "A semimodular lattice of finite length" defined in Birkhoff [1] coincides with an AC-lattice of finite length in our sense. While, "an semimodular atomic lattice" in Birkhoff [1] coincides with an M-symmetric atomistic lattice, which is an AC-lattice.

Exercise 8.1. Let a_i $(i = 1, \ldots, n)$ be finite elements of an AC-lattice L. Prove that $\{a_1, \ldots, a_n\}$ is a semi-orthogonal family if and only if

$$h(a_1 \vee \cdots \vee a_n) = h(a_1) + \cdots + h(a_n).$$

Exercise 8.2. Prove (8.17).

9. Finite-modular AC-lattices

Definition (9.1). In a lattice L, an element a is called a *modular element* when $(x, a)M$ for every $x \in L$. The elements 0, 1 and every atom, if they exist, are modular elements.

A lattice L with 0 is called *finite-modular* when every finite element of L is modular.

Lemma (9.2). *Let L be an AC-lattice. The following six statements are equivalent.*

(α) *L is finite-modular.*
(β) *If $h(b)=2$ in L then b is modular.*
(γ) *If $h(b)=2$ in L then $a \lessdot a \vee b$ implies $a \wedge b \lessdot b$.*
(δ) *If p and q are atoms and $p \leqq q \vee a$ in L $(a \neq 0)$ then there exists an atom $r \in L$ such that $p \leqq q \vee r$ and $r \leqq a$.*
(ε) *If p is an atom, b is a finite element and $p \leqq a \vee b$ in L $(a \neq 0, b \neq 0)$ then there exist atoms $q, r \in L$ such that $p \leqq q \vee r$, $q \leqq a$ and $r \leqq b$.*
(ζ) *If a is finite in L then $(a,x)M$ and $(x,a)M$ for every $x \in L$.*

Proof. The implications (ζ)⇒(α)⇒(β) are trivial. (β) implies (γ) by (7.5.3).

(γ)⇒(δ). Let $p \leqq q \vee a$. When $p=q$, any atom $r \leqq a$ may be used, and when $q \leqq a$, $r=p$ may be used. Hence we may assume $p \neq q$ and $q \nleqq a$. Since $a \vee (p \vee q) = a \vee q \gtrdot a$ and $h(p \vee q)=2$, by (γ) we have $a \wedge (p \vee q) \lessdot p \vee q$. Hence $r = a \wedge (p \vee q)$ is an atom. We have $q \neq r$ since $q \nleqq a$ and $r \leqq a$. Hence by the exchange property $r \leqq p \vee q$ implies $p \leqq q \vee r$. Thus (δ) holds.

(δ)⇒(ε). Let $p \leqq a \vee b$ and b be finite. There is a finite number of atoms $r_1, \ldots, r_n$ such that $b = r_1 \vee \cdots \vee r_n$. Since $p \leqq r_1 \vee \cdots \vee r_n \vee a$, by (δ) there exists an atom q_1 such that

$$p \leqq r_1 \vee q_1 \quad \text{and} \quad q_1 \leqq r_2 \vee \cdots \vee r_n \vee a.$$

Applying (δ) again, there exists an atom q_2 such that

$$q_1 \leqq r_2 \vee q_2 \quad \text{and} \quad q_2 \leqq r_3 \vee \cdots \vee r_n \vee a.$$

Continuing this process, lastly we get an atom q_n such that

$$q_{n-1} \leqq r_n \vee q_n \quad \text{and} \quad q_n \leqq a.$$

Putting $q = q_n$, we have

$$p \leqq r_1 \vee \cdots \vee r_n \vee q = q \vee b \quad \text{and} \quad q \leqq a.$$

By (δ) there exists an atom r such that

$$p \leqq q \vee r \quad \text{and} \quad r \leqq b.$$

(ε)⇒(ζ). We shall prove the following property which is equivalent to (ζ): If either a or b is finite then $(a,b)M$. Let $c \leqq b$. If p is an atom such that $p \leqq (c \vee a) \wedge b$, then since $p \leqq c \vee a$ and since either c or a is finite, it follows from (ε) that there exist atoms q and r such that

$$p \leqq q \vee r, \qquad q \leqq c \quad \text{and} \quad r \leqq a.$$

Since $(r,b)M$ by the covering property and since $q \leqq c \leqq b$, we have

$$p \leqq (q \vee r) \wedge b = q \vee (r \wedge b) \leqq c \vee (a \wedge b).$$

Since L is atomistic we have $(c \vee a) \wedge b \leqq c \vee (a \wedge b)$. Thus $(a,b)M$ holds. ▯

Theorem (9.3). *Let a and b be non-zero elements of a finite-modular AC-lattice L. Then $(a,b)M^*$ holds if and only if the following statement holds:*

(9.3.1) *If p is an atom such that $p \leqq a \vee b$ then there exist two atoms q and r such that $p \leqq q \vee r$, $q \leqq a$ and $r \leqq b$.*

Proof. (I) Assume that (9.3.1) holds. Let $c \geqq b$. If p is an atom such that $p \leqq c \wedge (a \vee b)$, then by (9.3.1) there exist atoms q and r such that $p \leqq q \vee r$, $q \leqq a$ and $r \leqq b$. When $p = r$ then $p \leqq b \leqq (c \wedge a) \vee b$. When $p \neq r$, by the exchange property we have $q \leqq p \vee r \leqq c \vee b = c$. Hence $q \leqq c \wedge a$, whence $p \leqq q \vee r \leqq (c \wedge a) \vee b$. Therefore $c \wedge (a \vee b) \leqq (c \wedge a) \vee b$, and $(a,b)M^*$ holds.

(II) Assume $(a,b)M^*$. Let $p \leqq a \vee b$ and put $c = b \vee p$. When $a \wedge c \nleqq b$, we have an atom q such that $q \leqq a \wedge c$ and $q \nleqq b$. Since $q \leqq c = p \vee b$, it follows from (δ) of (9.2) that there exists an atom r such that $q \leqq p \vee r$ and $r \leqq b$. Since $q \nleqq b$, we have $q \neq r$. Hence by the exchange property we have $p \leqq q \vee r$. Next, when $a \wedge c \leqq b$, it follows from $(a,b)M^*$ that

$$b = (c \wedge a) \vee b = c \wedge (a \vee b) \geqq p.$$

Hence the atom $r = p$ together with any atom $q \leqq a$ have the desired property. ▯

Corollary (9.4). *Let L be a finite-modular AC-lattice. If a is a finite element of L then $(a,x)M$, $(x,a)M$, $(a,x)M^*$ and $(x,a)M^*$ for every $x \in L$.*

Proof. $(a,x)M$ and $(x,a)M$ hold by (ζ) of (9.2). $(a,x)M^*$ and $(x,a)M^*$ hold by (ε) of (9.2) and (9.3). ▯

Theorem (9.5). *Let L be an AC-lattice. The following five statements are equivalent.*

(α) *L is finite-modular.*
(β) *L is M^*-symmetric.*
(γ) *If p is an atom of L then $(x,p)M^*$ for every $x \in L$.*
(δ) *$a \lessdot a \vee b$ implies $a \wedge b \lessdot b$ in L.*
(ε) *$a \lessdot a \vee b$ implies $(a,b)M$ in L.*

Proof. (α) implies (β) by (9.3). (β) implies (γ) since L satisfies (γ) of (7.6).

(γ)⇒(δ). If $a \lessdot a \vee b$, then since $b \nleqq a$, there exists an atom p such that $p \leqq b$ and $p \nleqq a$. Then $a < a \vee p \leqq a \vee b$, which implies $a \vee p = a \vee b$. By (γ) we have

$$a \wedge b \lessdot (b \wedge a) \vee p = b \wedge (a \vee p) = b \wedge (a \vee b) = b.$$

(δ) implies (ε) by (7.5.1), and (ε) implies (δ) by (7.5.3). Finally, (δ) implies (α) since (δ) implies (γ) of (9.2). □

Corollary (9.6). *If a finite-modular AC-lattice L with* 1 *is dual section semicomplemented, then L^* is also a finite-modular AC-lattice and then L is M-symmetric and M*-symmetric.*

Proof. L^* is atomistic by (8.19) and is M-symmetric by (9.5). Hence L^* is an AC-lattice. Since $a \wedge b \lessdot a$ implies $b \lessdot a \vee b$ in L by (7.10), L^* satisfies (δ) of (9.5). Hence L^* is finite-modular. L is M-symmetric since L^* is M*-symmetric. □

Remark (9.7). Let L be an atomistic lattice with the finite covering property (8.17.1). Since $\mathscr{I}(L)$ is an AC-lattice, the following eight statements are equivalent by (9.2) and (9.5).

(α) $\mathscr{I}(L)$ is modular.
(α′) If $a, b \in \mathscr{I}(L)$ and $h(b) = 2$ then $(a, b)M$.
(β) $\mathscr{I}(L)$ is M*-symmetric.
(γ) If $p \in \Omega(L)$ and $a \in \mathscr{I}(L)$ then $(a, p)M^*$.
(δ) If $a, b \in \mathscr{I}(L)$ then $a \lessdot a \vee b$ implies $a \wedge b \lessdot b$.
(δ′) If $a, b \in \mathscr{I}(L)$ and $h(b) = 2$ then $a \lessdot a \vee b$ implies $a \wedge b \lessdot b$.
(ε) If $p, q \in \Omega(L)$, $a \in \mathscr{I}(L)$ and $p \leqq q \vee a$ $(a \neq 0)$ then there exists $r \in \Omega(L)$ such that $p \leqq q \vee r$ and $r \leqq a$.
(ζ) If $p \in \Omega(L)$, $a, b \in \mathscr{I}(L)$ and $p \leqq a \vee b$ $(a \neq 0,\ b \neq 0)$ then there exist $q, r \in \Omega(L)$ such that $p \leqq q \vee r$, $q \leqq a$ and $r \leqq b$.

Exercise 9.1. Prove that if a and b are modular elements then so is $a \wedge b$.

Exercise 9.2. Prove that if a finite-modular AC-lattice L (without 1) is relatively complemented then L is M-symmetric.

Exercise 9.3. Let a be a finite element of an AC-lattice. Prove that if every element $b \leqq a$ with $h(b) = 2$ is modular then a is modular.

10. Distributivity and Perspectivity in Atomistic Lattices

Lemma (10.1). *Let L be a lattice with* 0. *An atom p of L is subperspective to* $a \in L$ *if and only if* $a \triangledown p$ *does not hold.*

Proof. If p is subperspective to a then by (6.2) $a \triangledown p$ does not hold. Conversely, if $a \triangledown p$ does not hold, then there exists $x \in L$ such that $(x \vee a) \wedge p > x \wedge p$. Since $(x \vee a) \wedge p > 0$, we have $p \leqq a \vee x$. Since $p > x \wedge p$, we have $p \wedge x = 0$. Hence p is subperspective to a. □

Lemma (10.2). *Let a and b be elements of an atomistic lattice L. The following four statements are equivalent.*

(α) $a \triangledown b$.
(β) $a \triangledown p$ *for every atom* $p \leqq b$.
(γ) *There exists no atom* $p \leqq b$ *which is subperspective to a.*
(δ) *There exists no non-zero element* $b_1 \leqq b$ *which is subperspective to a.*

Proof. Since L is SSC, the equivalence of (α) and (δ) follows from (6.3). The equivalence of (β) and (γ) follows from (10.1). The implication (δ)⇒(γ) is trivial. We shall prove the implication (γ)⇒(α). If $(x \vee a) \wedge b > x \wedge b$, then there exists an atom p such that $p \leqq (x \vee a) \wedge b$ and $p \nleqq x \wedge b$. Then $p \leqq a \vee x$, and since $p \leqq b$ we have $p \wedge x = p \wedge x \wedge b = 0$. This contradicts (γ). □

Lemma (10.3). *Let L be a SSC* lattice with* 0, *and let p and q be atoms of L. The following three statements are equivalent.*

(α) $p \sim q$.
(β) *q is subperspective to p.*
(γ) $p \triangledown q$ *does not hold.*

Proof. The equivalence of (β) and (γ) follows from (10.1), and the implication (α)⇒(β) is trivial.

(γ)⇒(α). If $p \triangledown q$ does not hold then by (4.18) there exists $x \in L$ such that $x < (x \vee p) \wedge (x \vee q)$. By (4.17.2) there exists $c \in L$ such that

$$1 > c \geqq x \quad \text{and} \quad c \vee \{(x \vee p) \wedge (x \vee q)\} = 1.$$

Then $c \vee p = c \vee x \vee p = 1$ and $c \vee q = c \vee x \vee q = 1$. If $c \wedge p \neq 0$, then $p \leqq c$, whence $c = c \vee p = 1$, a contradiction. Hence $c \wedge p = 0$. Similarly $c \wedge q = 0$. Therefore $p \sim_c q$.

Lemma (10.4). *Let L be an atomistic SSC* lattice with* 1, *and let p be an atom of L.*

(10.4.1) *If p is subperspective to* $a \in L$ *then there exists an atom q such that* $p \sim q \leqq a$.

(10.4.2) *If p is subprojective to* $a \in L$ *then there exist atoms* $q_0, q_1, \ldots, q_n$ *such that* $q_0 = p$, $q_n \leqq a$ *and* $q_{i-1} \sim q_i$ *for every* $i = 1, \ldots, n$.

(10.4.3) *If q is an atom of L and if* $p \approx q$ *then there exist atoms* $q_0, q_1, \ldots, q_n$ *such that* $q_0 = p$, $q_n = q$ *and* $q_{i-1} \sim q_i$ *for every* $i = 1, \ldots, n$.

Proof. If p is subperspective to a then by (10.1) $a \nabla p$ does not hold. By (4.19.1) $p \nabla a$ does not hold. Hence by (10.2) there exists an atom $q \leqq a$ which is subperspective to p. By (10.3) we have $p \sim q$. Thus (10.4.1) holds. (10.4.2) follows from (10.4.1), and (10.4.3) follows from (10.4.2). □

Theorem (10.5). *Let a and b be elements of an atomistic SSC* lattice L. The following three statements are equivalent.*

(α) $a \nabla b$.

(β) *There do not exist atoms p and q such that*

$$p \sim q, \quad p \leqq a \quad \text{and } q \leqq b.$$

(γ) *There do not exist non-zero elements* a_1 *and* b_1 *such that*

$$a_1 \sim b_1, \quad a_1 \leqq a \quad \text{and } b_1 \leqq b.$$

Proof. (α)⇒(γ). If $a_1 \sim b_1$, $a_1 \leqq a$ and $b_1 \leqq b$, then since $a_1 \nabla b_1$ by (α), we have $a_1 = b_1 = 0$ by (6.2).

The implication (γ)⇒(β) is trivial.

(β)⇒(α). If $a \nabla b$ does not hold, then by (10.2) there exists an atom $p \leqq b$ which is subperspective to a. By (10.4.1), there exists an atom q such that $p \sim q \leqq a$. This contradicts (β). □

Remark (10.6). A dual-atomistic lattice is a SSC* lattice by (7.3). Hence the result (10.4) and (10.5) are valid when L is an atomistic dual-atomistic lattice.

Lemma (10.7). *If L is a complete atomistic SSC* lattice, in particular if L is a complete atomistic dual-atomistic lattice, then L is a Z-lattice.*

Proof. Since L is both SSC and SSC*, L is a Z-lattice by (5.14). □

Lemma (10.8). *Any compactly atomistic lattice is a Z-lattice.*

Proof. By (7.13) a compactly atomistic lattice L is upper continuous, whence L is ∇-continuous by (5.5). Moreover, L is SSC. Hence L is a Z-lattice by (5.13). □

Theorem (10.9). *Let a and b be elements of a complete atomistic SSC* lattice. The following three statements are equivalent.*

(α) $e(a) \wedge e(b) = 0$.

(β) *There do not exist atoms p and q such that*

$$p \approx q, \quad p \leqq a \quad \text{and} \quad q \leqq b.$$

(γ) *There do not exist non-zero elements a_1 and b_1 such that*

$$a_1 \approx b_1, \quad a_1 \leqq a \quad \text{and} \quad b_1 \leqq b.$$

Proof. The implication (γ)⇒(β) is trivial. (α)⇒(γ) follows from (6.9).

(β)⇒(α). If $e(a) \wedge e(b) \neq 0$, then by (6.9) there exists a non-zero element $b_1 \leqq b$ which is subprojective to a. Take an atom q with $q \leqq b_1$. Since q is subprojective to a, by (10.4.2) there exists an atom p such that $q \approx p \leqq a$. Hence (β) does not hold. □

Corollary (10.10). *Let a and b be elements of a complete atomistic SSC* lattice L. If in L the perspectivity of atoms is transitive then the following four statements are equivalent.*

(α) $a \triangledown b$.

(β) $e(a) \wedge e(b) = 0$.

(γ) *There do not exist atoms p and q such that*

$$p \sim q, \quad p \leqq a \quad \text{and} \quad q \leqq b.$$

(δ) *There do not exist non-zero elements a_1 and b_1 such that*

$$a_1 \sim b_1, \quad a_1 \leqq a \quad \text{and} \quad b_1 \leqq b.$$

Proof. If the perspectivity of atoms is transitive, then it follows from (10.4.3) that $p \sim q$ is equivalent to $p \approx q$. Hence this corollary follows from (10.5) and (10.9). □

Lemma (10.11). *If p is an atom of a Z-lattice L, then the central cover $e(p)$ is an atom of the center $Z(L)$.*

Proof. Let p be an atom of L. If $z \in Z(L)$ with $0 < z \leqq e(p)$, then by (5.11.2) we have

$$e(z \wedge p) = z \wedge e(p) = z > 0.$$

Hence $z \wedge p \neq 0$, whence $p \leqq z$. Therefore $e(p) = z$. □

Lemma (10.12). *Let L be an atomistic Z-lattice. Then the center $Z(L)$ is atomistic. Moreover, if $\{z_\alpha; \alpha \in I\}$ is the set of all atoms of $Z(L)$, then the sublattice $L[0, z_\alpha]$ is irreducible for every $\alpha \in I$ and*

$$L = \dot{\bigcup}(L[0, z_\alpha]; \alpha \in I).$$

Proof. (I) For any $z \in Z(L)$, we have $z = \bigvee(p \in \Omega(L); p \leqq z)$ since L is atomistic. By (5.11.1) we have

$$\bigvee(e(p); p \in \Omega(L), p \leqq z) = e(z) = z.$$

By (10.11) this means that $Z(L)$ is atomistic.

(II) Let z_α be an atom of $Z(L)$. If z is a non-zero central element of the lattice $L[0, z_\alpha]$, then since $z_\alpha \in Z(L)$ it is easy to show that z is a central element of L. Since z_α is an atom of $Z(L)$, we have $z = z_\alpha$. Hence $L[0, z_\alpha]$ is irreducible.

(III) Let $\{z_\alpha; \alpha \in I\}$ be the set of all atoms of $Z(L)$. Then $z_\alpha \wedge z_\beta = 0$ for $\alpha \neq \beta$, and since $Z(L)$ is atomistic, we have $\bigvee(z_\alpha; \alpha \in I) = 1$. Hence it follows from (5.8) that

$$L = \dot{\bigcup}(L[0, z_\alpha]; \alpha \in I). \quad \square$$

Remark (10.13). Let L be an atomistic Z-lattice. We denote the atom space of $Z(L)$ by Ω_0. Since $Z(L)$ is a complete Boolean atomistic lattice, it is easy to show that $Z(L)$ is isomorphic to the lattice of all subsets of Ω_0, by the mapping $z \to \{x \in \Omega_0; x \leqq z\}$.

If moreover L is $\triangledown$-continuous, then it follows from (5.9) that there is a one-to-one correspondence between direct sum decompositions of L and decompositions of Ω_0.

Theorem (10.14). *Let L be a complete atomistic SSC* lattice.*

(10.14.1) *L is the direct sum of irreducible sublattices $\{L[0, z_\alpha]; \alpha \in I\}$ where $z_\alpha \in Z(L)$ for every $\alpha \in I$.*

(10.14.2) *Two atoms p and q of L are contained in the same summand $L[0, z_\alpha]$ if and only if $p \approx q$.*

(10.14.3) *L is irreducible if and only if any two atoms of L are projective.*

Proof. Since the projectivity is an equivalence relation, the atom space $\Omega(L)$ is decomposed into disjoint classes $\{P_\alpha; \alpha \in I\}$ such that two atoms p and q belong to the same class if and only if $p \approx q$. Put $z_\alpha = \bigvee(p; p \in P_\alpha)$. We shall show that z_α is an atom of $Z(L)$. Let $p \in P_\alpha$. By (6.8), the central cover $e(p)$ is equal to the join of all elements subprojective to p. Since L is atomistic, $e(p)$ is equal to the join of all atoms subprojective to p. By (10.4.2), an atom q is subprojective to p if and only if $q \approx p$. Hence we have $e(p) = z_\alpha$. By (10.11) z_α is an atom of $Z(L)$. Therefore (10.14.1) follows from (10.12). (10.14.2) is obvious by the above argument. (10.14.3) follows from (10.14.1) and (10.14.2). $\square$

EXERCISE 10.1. Let $q_1, \ldots, q_n$ and p be atoms of a lattice L with 0. Prove that if p is subperspective to $q_1 \vee \cdots \vee q_n$ (in particular if $p \leqq q_1 \vee \cdots \vee q_n$) then p is subperspective to one of $q_1, \ldots, q_n$.

EXERCISE 10.2. Let L be a SSC* lattice with 0, and let q_α $(\alpha \in I)$ be atoms of L such that $\bigvee(q_\alpha; \alpha \in I)$ exists. Prove that if an atom p is subperspective to $\bigvee(q_\alpha; \alpha \in I)$ then p is perspective to one of the atoms $\{q_\alpha; \alpha \in I\}$.

11. Perspectivity in AC-Lattices

Lemma (11.1). *Let p and q be atoms of a lattice L with the covering property. The following three statements are equivalent.*

(α) $p \sim q$.
(β) *q is subperspective to p.*
(γ) *$p \triangledown q$ does not hold.*

Proof. The equivalence of (β) and (γ) follows from (10.1), and the implication $(\alpha) \Rightarrow (\beta)$ is trivial.

$(\beta) \Rightarrow (\alpha)$. If $p \leqq q \vee x$ and $p \wedge x = 0$, then we have $q \wedge x = 0$, for otherwise $x = q \vee x \geqq p$, contrary to $p \wedge x = 0$. Moreover, by the covering property we have $p \vee x = q \vee x$. Hence $p \sim_x q$. □

Lemma (11.2). *Let p and q be different atoms of an AC-lattice L. If $p \sim_x q$ and $x \in \mathscr{J}(L)$, then there exist atoms $s_i \leqq x$, $i = 1, \dots, n$ such that $(p, s_1, \dots, s_n) \perp$, $(q, s_1, \dots, s_n) \perp$, $(p, q, s_1, \dots, s_{i-1}, s_{i+1}, \dots, s_n) \perp$ and*

$$p \vee s_1 \vee \cdots \vee s_n = q \vee s_1 \vee \cdots \vee s_n = p \vee q \vee s_1 \vee \cdots \vee s_{i-1} \vee s_{i+1} \vee \cdots \vee s_n$$

for every $i = 1, \dots, n$.

Proof. Since $q \leqq p \vee x$ and $x \in \mathscr{J}(L)$, there exist atoms $s_1, \dots, s_n$ such that

(1) $$q \leqq p \vee s_1 \vee \cdots \vee s_n \quad \text{and} \quad s_i \leqq x \qquad (i = 1, \dots, n).$$

Without loss of generality, we may assume that

(2) $$q \nleqq p \vee s_1 \vee \cdots \vee s_{i-1} \vee s_{i+1} \vee \cdots \vee s_n \quad \text{for every } i = 1, \dots, n,$$

since otherwise s_i may be deleted.

Then by the exchange property we have

$$p \vee q \vee s_1 \vee \cdots \vee s_{i-1} \vee s_{i+1} \vee \cdots \vee s_n = p \vee s_1 \vee \cdots \vee s_n.$$

Since $p \wedge x = 0$, we have $p \wedge s_1 = 0$. Moreover, we have

$$(p \vee s_1 \vee \cdots \vee s_{i-1}) \wedge s_i = 0 \quad \text{for } i = 2, \dots, n,$$

since otherwise $q \leqq p \vee s_1 \vee \cdots \vee s_n = p \vee s_1 \vee \cdots \vee s_{i-1} \vee s_{i+1} \vee \cdots \vee s_n$, contrary to (2). Hence by (8.12) and (2.5) we have $(p, s_1, \dots, s_n) \perp$. Since $(s_1 \vee \cdots \vee s_n) \wedge q \leqq x \wedge q = 0$, we have $(q, s_1, \dots, s_n) \perp$. Moreover, by (2) we have $(p, q, s_1, \dots, s_{i-1}, s_{i+1}, \dots, s_n) \perp$. By (1) and the exchange property we have

$$p \vee s_1 \vee \cdots \vee s_n = q \vee s_1 \vee \cdots \vee s_n. \quad □$$

Lemma (11.3). *Let* $\{a_1, \ldots, a_n\}$ *be a semi-orthogonal family in a* $\perp$*-symmetric lattice. If we put*

$$b_i = a_1 \vee \cdots \vee a_{i-1} \vee a_{i+1} \vee \cdots \vee a_n \qquad (i = 1, \ldots, n-1)$$

then $b_1 \wedge \cdots \wedge b_{n-1} = a_n$.

Proof. By (3.3) we have $b_1 \wedge b_2 = a_3 \vee \cdots \vee a_n$. Applying (3.3) successively, we get the disired equation. □

Theorem (11.4). *Let* p, q *and* r *be atoms of an AC-lattice. If* $p \sim q$ *and* $q \sim r$ *in* $\mathscr{J}(L)$ *then* $p \sim r$ *in* $\mathscr{J}(L)$.

Proof. We may assume that p, q and r are different atoms. By (11.2) there exist two semi-orthogonal families of atoms $\{s_1, \ldots, s_m\}$ and $\{t_1, \ldots, t_n\}$ such that, setting

$$x = s_1 \vee \cdots \vee s_m, \qquad y = t_1 \vee \cdots \vee t_n \quad \text{and}$$
$$x_i = s_1 \vee \cdots \vee s_{i-1} \vee s_{i+1} \vee \cdots \vee s_m \qquad (i = 1, \ldots, m),$$

(1) $p \vee x = q \vee x = p \vee q \vee x_i \ (i = 1, \ldots, m), \quad p \perp x, \quad q \perp x, \quad (p, q, x_i) \perp \ (i = 1, \ldots, m).$

(2) $$q \vee y = r \vee y, \quad q \perp y \quad \text{and } r \perp y.$$

Since $p \vee x \vee y = p \vee x \vee t_1 \vee \cdots \vee t_n$ where $p \perp x$, by deleting t_j if $(p \vee x \vee t_1 \vee \cdots \vee t_{j-1}) \wedge t_j \neq 0$, we get a subset $\{t'_1, \ldots, t'_k\}$ of $\{t_1, \ldots, t_n\}$ such that

$$p \vee x \vee y = p \vee x \vee t'_1 \vee \cdots \vee t'_k \quad \text{and } (p, x, t'_1, \ldots, t'_k) \perp.$$

Putting $y' = t'_1 \vee \cdots \vee t'_k$, we have

(3) $$p \vee x \vee y = p \vee x \vee y' \quad \text{and } (p, x, y') \perp.$$

Since $p \vee x = q \vee x$ and $q \perp x$, we have $(q, x, y') \perp$, whence

$$(q, s_1, \ldots, s_m, y') \perp.$$

By (11.3) we have

$$(x \vee y') \wedge (q \vee x_1 \vee y') \wedge \cdots \wedge (q \vee x_m \vee y') = y'.$$

We have $r \not\leqq y'$ since $r \wedge y' \leqq r \wedge y = 0$. Hence

$$\begin{aligned} &\text{either } r \wedge (x \vee y') = 0 \\ &\text{or} \qquad r \wedge (q \vee x_i \vee y') = 0 \quad \text{for some } i. \end{aligned}$$

Since by (1), (2), and (3) we have

$$r \leqq q \vee y \leqq p \vee x \vee y = p \vee x \vee y' = p \vee q \vee x_i \vee y',$$

in either case r is subperspective to p in $\mathscr{J}(L)$. Hence $p \sim r$ in $\mathscr{J}(L)$ by (11.1). □

Remark (11.5). Let L be an AC-lattice. We shall give a direct sum decomposition of $\mathcal{J}(L)$. By (11.4) the atom space of L is decomposed into disjoint classes $\{P_\alpha; \alpha\in I\}$ where p and q belong to the same class if and only if $p\sim q$ in $\mathcal{J}(L)$. Denote by S_α the set consisting of 0 and all finite elements which are joins of atoms contained in P_α. Let $a\in S_\alpha$, $b\in S_\beta$ and $\alpha\neq\beta$. Then

$$a=p_1\vee\cdots\vee p_m,\qquad p_i\in P_\alpha \quad\text{and}$$
$$b=q_1\vee\cdots\vee q_n,\qquad q_j\in P_\beta.$$

For any i and j, we have $p_i\triangledown q_j$ in $\mathcal{J}(L)$ since otherwise by (11.1) we have $p_i\sim q_j$ in $\mathcal{J}(L)$, a contradiction. Hence by (4.6.2) we have $a\triangledown q_j$ for every j. Since $\mathcal{J}(L)$ is relatively complemented, in $\mathcal{J}(L)$ the relation "$\triangledown$" is symmetric by Exercise 4.3. Hence we have $a\triangledown b$. Therefore $S_\alpha\subset S_\beta^{\triangledown}$ in $\mathcal{J}(L)$. Though $\mathcal{J}(L)$ is not complete in general, any $a\in\mathcal{J}(L)$ can be expressed in the form

$$a=\bigvee(a_\alpha;\alpha\in I),\qquad a_\alpha\in S_\alpha\quad\text{for every }\alpha\in I,$$

where $a_\alpha=0$ except for a finite number of α. Hence we may say that $\mathcal{J}(L)$ is a direct sum of $\{S_\alpha;\alpha\in I\}$.

Moreover, we remark that (11.4) and the above result are valid when L is an atomistic lattice with the finite covering property (8.17.1).

Lemma (11.6). *Let p and q be different atoms of an AC-lattice L.*

(11.6.1) *If $p\vee q$ contains a third atom then p and q are perspective.*

When L is moreover finite-modular, the converse of (11.6.1) *is also true, more precisely,*

(11.6.2) *If $p\sim_x q$ then $p\vee q$ contains a third atom r such that $r\leqq x$.*

Proof. (I) If r is an atom such that $r\leqq p\vee q$, $r\neq p$ and $r\neq q$, then by the exchange property we have

$$p\vee r=p\vee q=q\vee r.$$

Hence $p\sim_r q$.

(II) Assume that L is finite-modular and let $p\sim_x q$. Since $p\leqq q\vee x$, by (δ) of (9.2) there exists an atom r such that $p\leqq q\vee r$ and $r\leqq x$. By the exchange property we have $r\leqq p\vee q$. Moreover we have $r\neq p$ since $p\wedge r\leqq p\wedge x=0$, and similarly $r\neq q$. Thus (11.6.2) holds. □

Theorem (11.7). *Let p, q and r be atoms of a finite-modular AC-lattice L. If $p\sim q$ and $q\sim r$ then $p\sim r$.*

Proof. It follows from (11.6) that if $p \sim q$ and $p \neq q$ then there is an atom r such that $p \sim_r q$. Hence, $p \sim q$ in L if and only if $p \sim q$ in $\mathscr{J}(L)$. Therefore, this theorem follows from (11.4).

Remark (11.8). (11.7) can be proved from the following fact instead of (11.4).

(11.8.1) If p, q and r are different atoms of a finite-modular AC-lattice L and if each of $p \vee q$ and $q \vee r$ contains a third atom, then $p \vee r$ contains a third atom.

The proof of (11.8.1) is as follows. If $p \vee q$ (resp. $q \vee r$) contains a third atom s (resp. t) then by the exchange property

$$p \vee s = q \vee s = p \vee q \quad \text{and} \quad q \vee t = r \vee t = q \vee r.$$

Since $t \leqq q \vee r \leqq s \vee p \vee r$, by ($\delta$) of (9.2) there exists an atom u such that

$$t \leqq s \vee u \quad \text{and} \quad u \leqq p \vee r.$$

When $u \neq p$ and $u \neq r$, u is a third atom contained in $p \vee r$. When $u = p$, we have $r \leqq q \vee t \leqq q \vee s \vee u = p \vee q$. Hence by the exchange property, q is a third atom contained in $p \vee r$. When $u = r$, we have $q \leqq r \vee t \leqq r \vee s \vee u = r \vee s$. By the exchange property we have $s \leqq q \vee r$, whence $p \leqq q \vee s \leqq q \vee r$. By the exchange property q is a third atom contained in $p \vee r$.

Lemma (11.9). *Let a and b be elements of a finite-modular AC-lattice L. If a is finite and b is subperspective to a, then b is finite and $h(b) \leqq h(a)$.*

Proof. Let $b \leqq a \vee x$, $b \wedge x = 0$ and $h(a) = n$. There exist atoms $p_1, \ldots, p_n$ such that $a = p_1 \vee \cdots \vee p_n$. We put

$$c_i = p_1 \vee \cdots \vee p_i \vee x \quad \text{and} \quad b_i = b \wedge c_i \qquad (i = 0, 1, \ldots, n).$$

Then $b_0 = b \wedge x = 0$, $b_n = b \wedge (a \vee x) = b$ and

$$b_i \wedge c_{i-1} = b \wedge c_i \wedge c_{i-1} = b \wedge c_{i-1} = b_{i-1}.$$

Since $c_{i-1} \leqq b_i \vee c_{i-1} \leqq c_i = c_{i-1} \vee p_i$, we have either

$$b_i \vee c_{i-1} = c_{i-1} \quad \text{or} \quad b_i \vee c_{i-1} > c_{i-1}.$$

When $b_i \vee c_{i-1} = c_{i-1}$, since $b_i \leqq c_{i-1}$, we have $b_{i-1} = b_i \wedge c_{i-1} = b_i$. When $b_i \vee c_{i-1} > c_{i-1}$, we have $b_{i-1} = b_i \wedge c_{i-1} < b_i$ by (δ) of (9.5). Since $b_0 = 0$ and $b_n = b$, b is finite and $h(b) \leqq n$. ☐

Remark (11.10). Let a and b be two finite elements of an AC-lattice. If L is finite-modular, then $a \sim b$ implies $h(a) = h(b)$ by (11.9). But in general, $a \sim b$ does not imply $h(a) = h(b)$.

Definition (11.11). If two elements a and b of an AC-lattice L have bases $\{p_\alpha;\ \alpha \in I\}$ and $\{q_\alpha;\ \alpha \in I\}$ respectively such that $p_\alpha \sim q_\alpha$ for all

$\alpha \in I$, then we say that a and b are *pointwise perspective*, and write $a \stackrel{\ell}{\sim} b$. When $a=b=0$, *we write* $a \stackrel{\ell}{\sim} b$ *also.*

It is evident that if a and b are finite elements then $a \stackrel{\ell}{\sim} b$ implies $h(a)=h(b)$.

Lemma (11.12). *If a complete AC-lattice L is a Z-lattice then in L* $a \stackrel{\ell}{\sim} b$ *implies* $e(a)=e(b)$.

Proof. By (6.7.3), $p_\alpha \sim q_\alpha$ implies $e(p_\alpha)=e(q_\alpha)$. Hence by (5.11.1) we have

$$e(a) = \bigvee(e(p_\alpha); \alpha \in I) = \bigvee(e(q_\alpha); \alpha \in I) = e(b). \quad \square$$

Exercise 11.1. Prove that in an AC-lattice if p is an atom then $a \triangledown p$ implies $p \triangledown a$.

Exercise 11.2. Prove that if an AC-lattice L has the property (11.6.2) then L is finite-modular.

12. Completion by Cuts

Definition (12.1). Let L be a lattice. For any subset X of L, we denote by X^u the set of upper bounds of X and denote by X^l the set of lower bounds of X. The following properties are easily verified.

(12.1.1) For every X, X^u is a dual-ideal and X^l is an ideal of L.

(12.1.2) $\quad X \subset Y$ implies $X^u \supset Y^u$ and $X^l \supset Y^l$.

(12.1.3) $X \subset X^{ul}$, $\quad X \subset X^{lu}$, $\quad X^u = X^{ulu}$ and $X^l = X^{lul}$ for every X.

We denote the family $\{X \subset L; X = X^{ul}\}$ by $\bar{L}$. If $X_\alpha \in \bar{L}$ for every $\alpha \in I$ then the set-intersection $\bigcap(X_\alpha; \alpha \in I)$ belongs to $\bar{L}$, because we have $(\bigcap_\alpha X_\alpha)^{ul} \subset X_\beta^{ul} = X_\beta$ for every $\beta \in I$, which imply $(\bigcap_\alpha X_\alpha)^{ul} = \bigcap_\alpha X_\alpha$. Hence $\bar{L}$ forms a complete lattice by set-inclusion, where $\bigwedge_\alpha X_\alpha = \bigcap_\alpha X_\alpha$ and $\bigvee_\alpha X_\alpha = (\bigcup_\alpha X_\alpha)^{ul}$. The greatest element of $\bar{L}$ is the whole set L. If L has 0 then the least element of $\bar{L}$ is the set $\{0\}$, and otherwise the least element is the empty set.

Lemma (12.2). *Let L be a lattice.*

(12.2.1) *For every* $a \in L$, *the principal ideal* $J_a = \{x \in L, x \leqq a\}$ *belongs to* $\bar{L}$.

(12.2.2) *The mapping* $a \to J_a$ *is an isomorphism between L and the sub-lattice* $\{J_a; a \in L\}$ *of* $\bar{L}$.

(12.2.3) *If L is complete then L is isomorphic to* $\bar{L}$ *by the mapping* $a \to J_a$.

Proof. (I) Since $(J_a)^u = \{x \in L; x \geqq a\}$, we have $(J_a)^{ul} = J_a$.

(II) The mapping $a \to J_a$ is one-to-one, since $J_a = J_b$ implies $a = b$. Since $J_{a \wedge b}$ is the set-intersection of J_a and J_b, $J_{a \wedge b} = J_a \wedge J_b$ in $\bar{L}$. If $J_a \leqq X$ and $J_b \leqq X$ in $\bar{L}$ then for every $y \in X^u$ we have $a \leqq y$ and $b \leqq y$, whence $a \vee b \leqq y$. Hence $a \vee b \in X^{ul} = X$, and then $J_{a \vee b} \leqq X$. Therefore $J_{a \vee b} = J_a \vee J_b$ in $\bar{L}$. Thus $\{J_a; a \in L\}$ is a sublattice of $\bar{L}$ which is isomorphic to L.

(III) Let L be complete. For an arbitrary $X \in \bar{L}$, we put $a = \bigvee(x; x \in X)$. Evidently, $X \subset J_a$. If $y \in X^u$, then since $x \leqq y$ for every $x \in X$, we have $a \leqq y$. Hence $a \in X^{ul} = X$. Therefore $J_a = X$. Thus the mapping $a \to J_a$ is an isomorphism of L onto $\bar{L}$. □

Definition (12.3). Let L be a lattice. The complete lattice $\bar{L} = \{X \subset L; X = X^{ul}\}$ is called *completion by cuts* (or *normal completion*) of L. We remark that the mapping $a \to J_a$ preserves any existing joins and meets.

Lemma (12.4). *Let $\bar{L}$ be the completion by cuts of a lattice L. If X, $Y \in \bar{L}$ then $x \in X^u$ and $y \in Y^u$ imply $x \vee y \in (X \vee Y)^u$ and $x \wedge y \in (X \wedge Y)^u$.*

Proof. $x \wedge y \in (X \wedge Y)^u$ holds evidently, since $x, y \in (X \wedge Y)^u$. Moreover, since $X \vee Y = (X \cup Y)^{ul}$, we have

$$x \vee y \in (X \cup Y)^u = (X \cup Y)^{ulu} = (X \vee Y)^u. \quad \square$$

Lemma (12.5). *Let $\bar{L}$ be the completion by cuts of a lattice L. Then $(a, b)M$ in L if and only if $(J_a, J_b)M$ in $\bar{L}$.*

Proof. If $(J_a, J_b)M$ in $\bar{L}$ then $(a, b)M$ in L, since $x \to J_x$ is an isomorphism between L and a sublattice of $\bar{L}$ by (12.2.2).

Conversely, assume that $(a, b)M$ in L. Let $X \leqq J_b$ in $\bar{L}$. It suffices to show that $(X \vee J_a) \wedge J_b \leqq X \vee (J_a \wedge J_b)$. If $x \in \{X \vee (J_a \wedge J_b)\}^u = (X \vee J_{a \wedge b})^u$, then $x \in X^u$ and $a \wedge b \leqq x$. Since $X \leqq J_b$ implies $b \in X^u$, we have $x \wedge b \in X^u$. Hence by (12.4) we have

$$\{(x \wedge b) \vee a\} \wedge b \in \{(X \vee J_a) \wedge J_b\}^u.$$

By $(a, b)M$, we have

$$\{(x \wedge b) \vee a\} \wedge b = (x \wedge b) \vee (a \wedge b) \leqq x.$$

Hence $x \in \{(X \vee J_a) \wedge J_b\}^u$. Therefore $\{X \vee (J_a \wedge J_b)\}^u \subset \{(X \vee J_a) \wedge J_b\}^u$, which implies $X \vee (J_a \wedge J_b) \geqq (X \vee J_a) \wedge J_b$. □

Lemma (12.6). *If L is an atomistic lattice, then $\bar{L}$ is atomistic, and then an element X of $\bar{L}$ is an atom if and only if there exists an atom p of L such that $X = J_p$. Moreover $\mathscr{J}(\bar{L}) = \{J_a; a \in \mathscr{J}(L)\}$.*

Proof. (I) If p is an atom of L, then we have $\{0\} \lessdot J_p$ in $\bar{L}$, since $J_p = \{0, p\}$. Conversely, if $\{0\} \lessdot X$ in $\bar{L}$, then since X contains a non-zero element, it contains an atom p. Since $\{0\} < J_p \leqq X$, we have $X = J_p$.

(II) Let $X < Y$ in $\bar{L}$. Taking an element $a \in Y$ such that $a \notin X$, there exists $b \in X^u$ such that $a \nleqq b$ since $a \notin X^{ul}$. Since L is atomistic, there exists an atom p such that $p \leqq a$ and $p \nleqq b$. We have $J_p \leqq Y$ in $\bar{L}$ since $p \leqq a \in Y$. On the other hand, since $p \nleqq b \in X^u$, we have $p \notin X^{ul} = X$, whence $J_p \nleqq X$. Therefore $\bar{L}$ is atomistic by (7.2).

(III) If $X \in \mathscr{J}(\bar{L})$, then since X is the join of a finite number of atoms $J_{p_1}, \dots, J_{p_n}$, we have $X = J_a$ with $a = p_1 \vee \cdots \vee p_n \in \mathscr{J}(L)$. Conversely, if $a = p_1 \vee \cdots \vee p_n$ then $J_a = J_{p_1} \vee \cdots \vee J_{p_n} \in \mathscr{J}(\bar{L})$. □

Theorem (12.7). *If L is an AC-lattice, then its completion by cuts $\bar{L}$ is also an AC-lattice. If L is moreover finite-modular, so is $\bar{L}$.*

Proof. (I) Let L be an AC-lattice. $\bar{L}$ is atomistic by (12.6). We shall show that if J_p is an atom of $\bar{L}$, then $(J_p, X)M$ for every $X \in \bar{L}$. We may assume $J_p \nleqq X$. Since $p \notin X = X^{ul}$, there exists $a \in X^u$ such that $p \nleqq a$. Since p is an atom of L, by (7.6) we have $(p, a)M$. Hence $(J_p, J_a)M$ in $\bar{L}$ by (12.5). Since $J_p \wedge J_a = J_0$ and $X \leqq J_a$, it follows from (1.5.3) that $(J_p, X)M$. Hence $\bar{L}$ has the covering property by (7.6).

(II) Let L be a finite-modular AC-lattice. By (12.6), a finite element of $\bar{L}$ has the form J_a with $a \in \mathscr{J}(L)$. We need to show that $(X, J_a)M$ for every $X \in \bar{L}$. We may assume $J_a \nleqq X$. We shall prove that

(1) there exists $x \in X^u$ such that $a \wedge x \in X$.

Since $a \notin X = X^{ul}$, there exists $x_1 \in X^u$ such that $a \nleqq x_1$. Putting $a_1 = a \wedge x_1$, we have $a_1 < a$. When $a_1 \in X$, $x = x_1$ may be used. When $a_1 \notin X$, there exists $x_2 \in X^u$ such that $a_1 \nleqq x_2$. Putting $a_2 = a_1 \wedge x_2$, we have $a_2 < a_1$. When $a_2 \in X, x = x_1 \wedge x_2$ may be used, since $a \wedge x_1 \wedge x_2 = a_1 \wedge x_2 = a_2 \in X$. Since a is finite, continuing this process, finally we get $a_n = 0$ for some n. Then $x = x_1 \wedge \cdots \wedge x_n$ may be used, since $a \wedge x_1 \wedge \cdots \wedge x_n = a_n = 0 \in X$. Thus (1) has been proved.

By (1) we have $X \leqq J_x$ and $J_x \wedge J_a \leqq X$. Since $(J_x, J_a)M$ by (12.5), we have $(X, J_a)M$ by (1.5.2). □

Exercise 12.1. Prove that if $a \lessdot b$ in a lattice L then $J_a \lessdot J_b$ in $\bar{L}$.

Exercise 12.2. Prove that if $a < b$ in a lattice L then $\overline{L[a, b]}$ and $\bar{L}[J_a, J_b]$ are isomorphic.

Exercise 12.3. Prove that a lattice L with 0 is semicomplemented (resp. SSC) if and only if $\bar{L}$ is semicomplemented (resp. SSC).

Problem 2. Is there an AC-lattice which is not M-symmetric?

PROBLEM 3. Is there an AC-lattice where $a \triangledown b$ does not imply $b \triangledown a$?

PROBLEM 4. Is the completion by cuts of an M-symmetric lattice M-symmetric?

References for Chapter II

For Section 7: G. Birkhoff [1] (Chapter VIII), F. Maeda [1], D. Sachs [1].

For Section 8: G. Birkhoff [1] (Chapter IV), M. D. MacLaren [1], F. Maeda [1].

For Section 9: S. Maeda [3], M. F. Janowitz [5].

For Section 10: M. F. Janowitz [4], S. Maeda [6].

For Section 11: S. Maeda [6], U. Sasaki and S. Fujiwara [1].

For Section 12: G. Birkhoff [1] (Chapter V), M. F. Janowitz [4] and [5].

Chapter III

Matroid Lattices

13. Perspectivity and Irreducible Decompositions of Matroid Lattices

In (7.16), a matroid lattice is defined as a compactly atomistic M-symmetric lattice, and it may be defined as an upper continuous AC-lattice. It was shown in (7.10) and (7.15) that in a compactly atomistic lattice (α) the property of being M-symmetric, (β) the covering property, and (γ) the exchange property are equivalent. In Dubeil-Jacotin, Leisieur and Croisot [1] a compactly atomistic lattice with (β) is called a geometric lattice, and in MacLane [1] a compactly atomistic lattice with (γ) is called an exchange lattice.

Remark (13.1). A matroid lattice is left complemented by (7.15). Since it is relatively complemented, it is dual-atomistic by (8.19). Moreover it is a Z-lattice by (10.8).

Remark (13.2). If $a<b$ in a matroid lattice L, then $L[a,b]$ is a matroid sublattice of L. This is so, because $L[a,b]$ is an AC-lattice by (8.18), and since L is upper continuous, so is $L[a,b]$.

Theorem (13.3). *Let p, q and r be atoms of a matroid lattice L. If $p\sim q$ and $q\sim r$ then $p\sim r$.*

Proof. If $p\sim_x q$, then since $p\leqq q\vee x$ and since L is compactly atomistic, there exists a finite element y such that $p\leqq q\vee y$ and $y\leqq x$. Since $p\wedge y\leqq p\wedge x=0$, we have $p\sim_y q$ by (11.1). Therefore $p\sim q$ in L if and only if $p\sim q$ in $\mathscr{J}(L)$. Hence this theorem follows from (11.4). ▯

Corollary (13.4). *If p and q are atoms of a matroid lattice L, then $p\approx q$ implies $p\sim q$.*

Proof. L is SSC* since it is relatively complemented. Hence by (10.4.3) this corollary follows from (13.3). ▯

Theorem (13.5). *Let a and b be elements of a matroid lattice. The following four statements are equivalent.*

(α) $a \triangledown b$.
(β) $e(a) \wedge e(b) = 0$.
(γ) *There do not exist atoms p and q such that*

$$p \sim q, \quad p \leqq a \quad \text{and } q \leqq b.$$

(δ) *There do not exist non-zero elements a_1 and b_1 such that*

$$a_1 \sim b_1, \quad a_1 \leqq a \quad \text{and } b_1 \leqq b.$$

Proof. This follows from (10.10) and (13.3). □

Theorem (13.6). *Let L be a matroid lattice.*

(13.6.1) *L is the direct sum of irreducible sublattices $\{L[0, z_\alpha]; \alpha \in I\}$ where $z_\alpha \in Z(L)$ for every $\alpha \in I$.*
(13.6.2) *Two atoms of L are contained in the same summand $L[0, z_\alpha]$ if and only if they are perspective.*
(13.6.3) *L is irreducible if and only if any two atoms of L are perspective.*

Proof. This follows from (10.14) and (13.4). □

Theorem (13.7). (Comparability theorem). *Let a and b be elements of a matroid lattice L. There exist elements a', a'', b' and b'' such that*

$$a = a' \vee a'', \quad a' \perp a'', \quad b = b' \vee b'', \quad b' \perp b'',$$

$$a' \stackrel{p}{\sim} b', \quad \text{and } e(a'') \wedge e(b'') = 0.$$

Moreover $e(a') = e(b') = e(a) \wedge e(b)$.

Proof. Let T be the set of all pairs (p, q) of atoms such that $p \leqq a$, $q \leqq b$, and $p \sim q$. If T is empty, then since $e(a) \wedge e(b) = 0$ by (13.5), we may put $a' = b' = 0$, $a'' = a$ and $b'' = b$. Assume that T is non-empty. Denote by Φ the family of all subsets S of T satisfying the following condition: Both $\{p; (p, q) \in S\}$ and $\{q; (p, q) \in S\}$ are semi-orthogonal families. By (2.3) a subset S of T belongs to Φ if and only if all finite subsets of S belong to Φ. Hence any chain in Φ has an upper bound in Φ. Therefore, by Zorn's lemma, there exists a maximal set S^* in Φ. Put

$$a' = \bigvee(p; (p, q) \in S^*) \quad \text{and } b' = \bigvee(q; (p, q) \in S^*).$$

By (11.11) we have $a' \stackrel{p}{\sim} b'$. Since L is relatively semi-orthocomplemented by (13.1) and (3.9), there exist a'' and b'' such that $a = a' \vee a''$, $a' \perp a''$, $b = b' \vee b''$ and $b' \perp b''$. If $e(a'') \wedge e(b'') \neq 0$, then by (13.5) there exist atoms p_1 and q_1 such that $p_1 \leqq a''$, $q_1 \leqq b''$ and $p_1 \sim q_1$. Since $(p_1, q_1) \in T$ and since $a' \perp p_1$ and $b' \perp q_1$, the set S^* with (p_1, q_1) adjoined belongs to Φ. This contradicts the maximality of S^*. Hence we have $e(a'') \wedge e(b'') = 0$.

By (11.12) and (5.11.1) we have

$$e(a')=e(b'), \quad e(a)=e(a')\vee e(a'') \quad \text{and} \quad e(b)=e(b')\vee e(b'').$$

Hence

$$\begin{aligned} e(a)\wedge e(b) &= (e(a')\vee e(a''))\wedge(e(b')\vee e(b'')) \\ &= e(a')\vee(e(a'')\wedge e(b''))=e(a'). \quad \square \end{aligned}$$

EXERCISE 13.1. Let L be an irreducible matroid lattice. Prove that if a is a modular element of L then $L[0,a]$ is an irreducible matroid lattice.

EXERCISE 13.2. Let $E(S)$ be the family of all equivalence relations on a given set S. We define $E_1 \leqq E_2$ in $E(S)$ when $x\equiv y(E_1)$ implies $x\equiv y(E_2)$. Prove that $E(S)$ is a matroid lattice. ($E(S)$ is sometimes called a partition lattice. Cf. Sachs [1].)

14. Modularity in Matroid Lattices

Theorem (14.1). *Let L be a matroid lattice. The following six statements are equivalent.*

(α) *L is modular.*

(β) *L is finite-modular.*

(γ) $\mathscr{I}(L)$ *is modular.*

(δ) *L is M*-symmetric.*

(ε) $\mathscr{I}(L)$ *is M*-symmetric.*

(ζ) *L has the dual covering property.*

Proof. (I) The implications $(\alpha)\Rightarrow(\beta)\Rightarrow(\gamma)$ are trivial. We shall prove $(\gamma)\Rightarrow(\alpha)$. Let a and b be arbitrary elements of L, and let $c\leqq b$. If p is an atom such that $p\leqq(c\vee a)\wedge b$, then since L is compactly atomistic, there exist two finite elements c_1 and a_1 such that $p\leqq c_1\vee a_1$, $c_1\leqq c$ and $a_1\leqq a$. Since $(a_1,c_1\vee p)M$ in $\mathscr{I}(L)$ by (γ), we have

$$p\leqq(c_1\vee a_1)\wedge(c_1\vee p)=c_1\vee\{a_1\wedge(c_1\vee p)\}\leqq c\vee(a\wedge b).$$

Therefore $(c\vee a)\wedge b\leqq c\vee(a\wedge b)$, whence $(a,b)M$.

(II) (β) and (δ) are equivalent by (9.5), and (γ) and (ε) are equivalent by (9.7). (δ) implies (ζ) by (7.7).

We shall prove $(\zeta)\Rightarrow(\beta)$. If L^* has the covering property, then since L^* is atomistic by (13.1), it follows from (7.10) that

$$a\wedge b \lessdot a \quad \text{implies} \quad b\lessdot a\vee b \quad \text{in } L^*.$$

Hence L satisfies (δ) of (9.5), whence L is finite-modular. $\square$

Definition (14.2). In a matroid lattice an element a is called a *line* (resp. a *plane*) when $h(a)=2$ (resp. $h(a)=3$). It follows from (9.2) and (14.1) that a matroid lattice L is modular if every line is a modular element in L.

Definition (14.3). A lattice L with 0 is called *strongly planar* when L satisfies the following condition:

(14.3.1) If p, q and r are atoms and if $p \leqq q \vee a$ and $r \leqq a$ then there exists an atom s such that

$$p \leqq q \vee r \vee s \quad \text{and} \quad s \leqq a.$$

This condition is weaker than (δ) of (9.2).

Lemma (14.4). *An AC-lattice L is strongly planar if and only if the sublattice $L[p, \rightarrow]$ is finite-modular for every atom p of L.*

Proof. (I) Let L be strongly planar and let p be an atom of L. By (8.18) the sublattice $L[p, \rightarrow]$ is an AC-lattice. We shall show that $L[p, \rightarrow]$ satisfies (δ) of (9.2). Let l and k be atoms of $L[p, \rightarrow]$ and $l \leqq k \vee a$ in $L[p, \rightarrow]$. By (8.18) there exist atoms q and r of L such that $p \neq q$, $l = p \vee q$, $p \neq r$ and $k = p \vee r$. Then

$$q \leqq l \leqq k \vee a = p \vee r \vee a = r \vee a \quad \text{and} \quad p \leqq a.$$

Since L is strongly planar, there exists an atom s such that

$$q \leqq r \vee p \vee s \quad \text{and} \quad s \leqq a.$$

When $s \neq p$, putting $d = p \vee s$, d is an atom of $L[p, \rightarrow]$ and we have

$$l = p \vee q \leqq r \vee p \vee s = k \vee d \quad \text{and} \quad d \leqq a.$$

When $s = p$, taking any atom d of $L[p, \rightarrow]$ with $d \leqq a$, we have

$$l \leqq r \vee p \vee s = r \vee p = k \leqq k \vee d.$$

Thus $L[p, \rightarrow]$ satisfies (δ) of (9.2), whence it is finite-modular.

(II) Assume that $L[p, \rightarrow]$ is finite-modular for any atom p. We shall show (14.3.1). Let p, q and r be atoms and let $p \leqq q \vee a$ and $r \leqq a$. When $p = r$ or $q = r$, we have $p \leqq a$. Hence $s = p$ may be used. When $p \neq r$ and $q \neq r$, then $p \vee r$ and $q \vee r$ are atoms of $L[r, \rightarrow]$ and $p \vee r \leqq (q \vee r) \vee a$ in $L[r, \rightarrow]$. Since $L[r, \rightarrow]$ is finite-modular, by (δ) of (9.2) there exists an atom l of $L[r, \rightarrow]$ such that

$$p \vee r \leqq (q \vee r) \vee l \quad \text{and} \quad l \leqq a.$$

Taking an atom s such that $l = r \vee s$, we have

$$p \leqq q \vee r \vee s \quad \text{and} \quad s \leqq a. \quad \square$$

Theorem (14.5). *A matroid lattice L is weakly modular if and only ifand only if L is strongly planar.*

Proof. For any atom p of L, the sublattice $L[p,1]$ is a matroid lattice by (13.2). Hence, by (14.1) and (14.4), L is strongly planar if and only if $L[p,1]$ is modular for every p. If L is weakly modular then evidently $L[p,1]$ is modular. Conversely, if $L[p,1]$ is modular for every p, then for two elements a and b such that $a \wedge b \neq 0$ taking an atom p with $p \leqq a \wedge b$, we have $(a,b)M$ in $L[p,1]$. Hence $(a,b)M$ in L by (1.4). Thus L is weakly modular. ▯

Theorem (14.6). *Let L be a matroid lattice. L is modular if and only if L satisfies the following condition:*

(14.6.1) *If p and q are different atoms such that $p \sim_x q$ then there exists an atom r such that $p \sim_r q$ and $r \leqq x$.*

L is weakly modular if and only if L satisfies the following condition:

(14.6.2) *If p and q are different atoms such that $p \sim_x q$ and if r is an atom with $r \leqq x$ then there exists either an atom s such that $p \sim_s q$ and $s \leqq x$ or a line l such that $p \sim_l q$ and $r < l \leqq x$.*

Proof. (I) Let L be modular. If $p \sim_x q$ then by (11.6.2) $p \vee q$ contains a third atom r with $r \leqq x$. Then $p \sim_r q$ by the exchange property.

Conversely assume that L satisfies (14.6.1). We shall show that (δ) in (9.2) holds in L. Let $p \leqq q \vee a$. When $p \leqq a$, $r = p$ may be used. When $p \nleqq a$, then $p \sim_a q$ by (11.1). Hence by (14.6.1) there exists an atom r such that $p \sim_r q$ and $r \leqq a$. Hence (δ) in (9.2) holds in L, which means that L is finite-modular. Therefore L is modular by (14.1).

(II) To prove the second statement, by (14.5) it suffices to show the equivalence of (14.3.1) and (14.6.2).

(14.3.1)$\Rightarrow$(14.6.2). Let $p \sim_x q$ and $r \leqq x$. By (14.3.1) there exists an atom s such that

$$p \leqq q \vee r \vee s \quad \text{and} \quad s \leqq x.$$

Then $p \wedge (r \vee s) \leqq p \wedge x = 0$. If $r \neq s$, then putting $l = r \vee s$, we have $p \sim_l q$ by (11.1). If $r = s$ then we have $p \sim_s q$.

(14.6.2)$\Rightarrow$(14.3.1). Let $p \leqq q \vee a$ and $r \leqq a$. When $p \leqq a$, $s = p$ may be used. When $p \nleqq a$, then since $p \sim_a q$ by (11.1), it follows from (14.6.2) that there exists either an atom s such that $p \sim_s q$ and $s \leqq a$ or a line l such that $p \sim_l q$ and $r < l \leqq a$. In the latter case there exists an atom s such that $r \vee s = l$. Hence in either case

$$p \leqq q \vee r \vee s \quad \text{and} \quad s \leqq a. \quad ▯$$

Theorem (14.7). *If a weakly modular lattice L with* 1 *is not modular then L is irreducible. If a weakly modular matroid lattice L is not modular then any two atoms of L are perspective.*

Proof. (I) Assume that L is a weakly modular lattice with 1 and that L is not modular. Let z be a central element of L, and z' be the complement of z. Putting $L_1 = L[0,z]$ and $L_2 = L[0,z']$, we have $L \cong L_1 L_2$. It follows from (1.18) that either both L_1 and L_2 are modular or one of L_1 and L_2 consists of only a zero element. But the former case does not occur since L is not modular. Hence $z=0$ or $z'=0$, which implies that z is equal to 0 or 1. Thus L is irreducible.

(II) If moreover L is a matroid lattice, then by (13.6.3) any two atoms of L are perspective. □

EXERCISE 14.1. Prove that a compactly atomistic lattice L is modular if it satisfies (δ) of (9.2) and the following condition:

$$\text{If } p,q,r \in \Omega(L), \quad p \neq q \text{ and } p \leqq q \vee r \text{ then } r \leqq p \vee q.$$

EXERCISE 14.2. Let L be a strongly planar compactly atomistic lattice. Prove that if p and q are atoms of L and if $p \leqq a \vee b$ and $q \leqq b$ in L then there exists an atom $r \in L$ such that $p \leqq a \vee q \vee r$ and $r \leqq b$.

Prove that if a and b are non-zero elements and p is an atom with $p \leqq a \vee b$ then there exist atoms q, r, s and t such that

$$p \leqq q \vee r \vee s \vee r; \quad q,r \leqq a; \quad s,t \leqq b.$$

EXERCISE 14.3. Let p, q and r be atoms of a strongly planar AC-lattice. Prove that if $p \sim q$ and $q \sim r$ then $p \sim r$.

EXERCISE 14.4. Prove that the completion by cuts of a strongly planar AC-lattice is strongly planar.

EXERCISE 14.5. Prove that an AC-lattice L is strongly planar if and only if $a \wedge b \neq 0$ and $(a,b)M^*$ together imply $(b,a)M^*$ in L.

EXERCISE 14.6. Prove that an AC-lattice of length 2 is always modular and that an AC-lattice of length 3 is always weakly modular.

15. Atom Spaces of Atomistic Lattices

Definition (15.1). Let L be an atomistic lattice. A subset ω of the atom space $\Omega(L)$ of L is called a *subspace* when ω satisfies the following condition:

(15.1.1) If $p \in \Omega(L)$, $q_i \in \omega$ $(i=1,\dots,n)$ and $p \leqq q_1 \vee \cdots \vee q_n$, then $p \in \omega$.

The empty set, the whole set and singleton sets are subspaces.

Remark (15.2). Let L be an atomistic lattice. If L satisfies (δ) of (9.2) (or more generally (ε) of (9.7)) then it is easy to show that (15.1.1) is equivalent to the following condition:

(15.2.1) If $p\in\Omega(L)$, $q_i\in\omega$ $(i=1,2)$ and $p\leqq q_1\vee q_2$ then $p\in\omega$.

If L is strongly planar then (15.1.1) is equivalent to the following condition:

(15.2.2) If $p\in\Omega(L)$, $q_i\in\omega$ $(i=1,2,3)$ and $p\leqq q_1\vee q_2\vee q_3$ then $p\in\omega$.

L may be called *planar* when (15.1.1) and (15.2.2) are equivalent (cf. Jónsson [1]).

Lemma (15.3). *Let S be a set and let Φ be a family of subsets of S. Φ forms an upper continuous lattice ordered by set-inclusion if Φ satisfies the following three conditions:*

(15.3.1) $$S\in\Phi.$$

(15.3.2) *If $A_\alpha\in\Phi$ for every $\alpha\in I$ then $\bigcap(A_\alpha;\alpha\in I)\in\Phi$.*

(15.3.3) *If $\{A_\delta;\delta\in D\}$ is an increasing family* (D is a directed set and $\delta_1\leqq\delta_2$ implies $A_{\delta_1}\subset A_{\delta_2}$) *and if $A_\delta\in\Phi$ for every $\delta\in D$, then $\bigcup(A_\delta;\delta\in D)\in\Phi$.*

Proof. By (15.3.1) and (15.3.2), the meet of elements A_α of Φ always exists, and hence Φ forms a complete lattice where $\bigwedge_\alpha A_\alpha=\bigcap_\alpha A_\alpha$. If $A_\delta\uparrow A$ in the lattice Φ, then $A=\bigcup_\delta A_\delta$ by (15.3.3). Hence for any $B\in\Phi$ we have $A\cap B=\bigcup_\delta(A_\delta\cap B)$. This means $A_\delta\wedge B\uparrow A\wedge B$ in Φ. Therefore Φ is upper continuous. ▯

Remark (15.4). The set $I(L)$ of all ideals of a lattice L forms an upper continuous lattice ordered by set-inclusion, since $I(L)$ satisfies the three conditions in (15.3). It is evident that the completion by cuts $\bar{L}$ is a subset of $I(L)$. But, in general, $\bar{L}$ is not a sublattice of $I(L)$ and the mapping $a\to J_a$ of L into $I(L)$ is not an isomorphism, though this mapping preserves any existing meet.

Theorem (15.5). *Let L be an atomistic lattice. The set $L(\Omega(L))$ of all subspaces of $\Omega(L)$ forms a compactly atomistic lattice ordered by set-inclusion. For any $a\in L$, the set $\omega(a)=\{p\in\Omega(L);p\leqq a\}$ is a subspace of $\Omega(L)$. The mapping $a\to\omega(a)$ is a one-to-one, order-preserving mapping of L into the compactly atomistic lattice $L(\Omega(L))$. This mapping has the following properties.*

(15.5.1) $$\omega(0)=0, \textit{ and } \omega(1)=1 \text{ (if } L \text{ has } 1).$$

(15.5.2) $$\omega(\textstyle\bigwedge_\alpha a_\alpha)=\bigwedge_\alpha\omega(a_\alpha) \text{ (if } \bigwedge_\alpha a_\alpha \text{ exists)}.$$

(15.5.3) $\omega(a \vee b) \geqq \omega(a) \vee \omega(b)$ *for every a and b, and equality holds if both a and b are finite.*

(15.5.4) $\mathscr{J}(L)$ *and* $\mathscr{J}(L(\Omega(L)))$ *are isomorphic by the mapping* $a \rightarrow \omega(a)$.

If L is compactly atomistic then L is isomorphic to $L(\Omega(L))$ *by the mapping* $a \rightarrow \omega(a)$.

Proof. (I) It is easily seen that $L(\Omega(L))$ satisfies the three conditions in (15.3). Hence $L(\Omega(L))$ forms an upper continuous lattice. Since any singleton set is an atom of $L(\Omega(L))$, $L(\Omega(L))$ is atomistic. Thus $L(\Omega(L))$ is a compactly atomistic lattice. It is easily seen that for $\omega_1, \omega_2 \in L(\Omega(L))$

(1)

$$\omega_1 \vee \omega_2 = \{p \in \Omega(L);\ p \leqq q_1 \vee \cdots \vee q_m \vee r_1 \vee \cdots \vee r_n,\ q_i \in \omega_1 \text{ and } r_j \in \omega_2\}.$$

(II) Evidently, $\omega(a)$ is a subspace of $\Omega(L)$ for every $a \in L$ and the mapping $a \rightarrow \omega(a)$ of L into $L(\Omega(L))$ is order-preserving. This mapping is one-to-one since L is atomistic. (15.5.1) and (15.5.2) are obvious. The inequality in (15.5.3) is obvious and by (1) equality holds if $a, b \in \mathscr{J}(L)$. It is easy to show that $\omega \in L(\Omega(L))$ is an atom if and only if $\omega = \omega(p)$ for some atom p of L. Hence (15.5.4) holds by (15.5.3).

(III) Let L be compactly atomistic. We shall show that the mapping $a \rightarrow \omega(a)$ is onto. Let $\omega \in L(\Omega(L))$, and put $a = \bigvee(q; q \in \omega)$. Evidently $\omega \subset \omega(a)$. If $p \in \omega(a)$ then $p \leqq \bigvee(q; q \in \omega)$. Since L is compactly atomistic, there exist $q_1, \ldots, q_n \in \omega$ such that $p \leqq q_1 \vee \cdots \vee q_n$. Hence $p \in \omega$. Therefore $\omega = \omega(a)$. Since the mapping $a \rightarrow \omega(a)$ is one-to-one, order-preserving and onto, it is an isomorphism. □

Remark (15.6). Let L be an atomistic lattice and let $\bar{L}$ be its completion by cuts. By (12.6) it is easy to verify that $L(\Omega(\bar{L}))$ is isomorphic to $L(\Omega(L))$.

Theorem (15.7). *If L is an AC-lattice, then the lattice* $L(\Omega(L))$ *of all subspaces of* $\Omega(L)$ *is a matroid lattice and it is isomorphic to the lattice of all ideals of* $\mathscr{J}(L)$.

Proof. (I) Let L be an AC-lattice. To prove that $L(\Omega(L))$ is a matroid lattice, it suffices to show that $L(\Omega(L))$ has the covering property. Let $\omega(p) \nleqq \omega$ in $L(\Omega(L))$. If $\omega < \omega_1 \leqq \omega \vee \omega(p)$, then taking an atom $q \in \omega_1$ with $q \notin \omega$, there exist $r_1, \ldots, r_n \in \omega$ such that $q \leqq r_1 \vee \cdots \vee r_n \vee p$. Since $q \notin \omega$ we have $q \nleqq r_1 \vee \cdots \vee r_n$. Hence by the exchange property we have $p \leqq r_1 \vee \cdots \vee r_n \vee q$, whence $p \in \omega_1$. Therefore $\omega_1 = \omega \vee \omega(p)$.

(II) By (8.8), $\mathscr{J}(L)$ is a sublattice of L. For any $\omega \in L(\Omega(L))$, let $\varphi(\omega)$ be the ideal of $\mathscr{J}(L)$ generated by all atoms in ω. For any ideal J of $\mathscr{J}(L)$, put $\psi(J) = J \cap \Omega(L)$. Evidently $\omega \subset \varphi(\omega) \cap \Omega(L) = \psi(\varphi(\omega))$. Conversely, if $p \in \varphi(\omega) \cap \Omega(L)$, then by the definition of $\varphi(\omega)$ there exist

$q_1, \ldots, q_n \in \omega$ such that $p \leqq q_1 \vee \cdots \vee q_n$. Since ω is a subspace we have $p \in \omega$. Hence $\omega = \psi(\varphi(\omega))$.

Next, it is evident that $\varphi(\psi(J)) = \varphi(J \cap \Omega(L)) \subset J$. Since any element of J is the join of a finite number of atoms in J, we have $\varphi(\psi(J)) = J$. Therefore φ and ψ are mutually inverse one-to-one mappings between $L(\Omega(L))$ and the set of ideals of $\mathscr{J}(L)$. Since they preserve the order, they are isomorphisms. □

We remark that this theorem is valid when L is an atomistic lattice with the finite covering property (8.17.1).

Corollary (15.8). *If L is a matroid lattice then L is isomorphic to the lattice of all ideals of* $\mathscr{J}(L)$.

Proof. Since L is compactly atomistic, L is isomorphic to $L(\Omega(L))$ by (15.5). Hence this corollary follows from (15.7). □

Remark (15.9). The lattice $I(L)$ of all ideals of an AC-lattice L is not necessarily atomistic. In fact, an element J of $I(L)$ is an atom if and only if $J = J_p$ for some $p \in \Omega(L)$. Hence, if $a \notin \mathscr{J}(L)$, then in $I(L)$ the join of atoms contained in J_a is equal to the ideal $J_a \cap \mathscr{J}(L)$ which is not equal to J_a.

Theorem (15.10). *If L is a strongly planar AC-lattice, then* $L(\Omega(L))$ *is a weakly modular matroid lattice. If L is a finite-modular AC-lattice, then* $L(\Omega(L))$ *is a modular matroid lattice and then it holds that*

(15.10.1) $\omega_1 \vee \omega_2 = \{p \in \Omega(L);\ p \leqq q \vee r,\ q \in \omega_1,\ r \in \omega_2\}$, and

(15.10.2) $\omega(a \vee b) = \omega(a) \vee \omega(b)$ *if and only if* $(a, b) M^*$ *in L.*

Proof. (I) Let L be a strongly planar AC-lattice. We shall show that $L(\Omega(L))$ is strongly planar. Let

$$\omega(p) \leqq \omega(q) \vee \omega \quad \text{and} \quad \omega(r) \leqq \omega \quad \text{in } L(\Omega(L)).$$

Then there exist $s_1, \ldots, s_n \in \omega$ such that $p \leqq q \vee s_1 \vee \cdots \vee s_n$. Since L is strongly planar, there exists $s \in \Omega(L)$ such that

$$p \leqq q \vee r \vee s \quad \text{and} \quad s \leqq s_1 \vee \cdots \vee s_n \vee r.$$

Then $\omega(p) \leqq \omega(q \vee r \vee s) = \omega(q) \vee \omega(r) \vee \omega(s)$ and $\omega(s) \leqq \omega$. Hence $L(\Omega(L))$ is strongly planar, whence it is weakly modular by (14.5).

(II) If L is a finite-modular AC-lattice, then since $\mathscr{J}(L)$ is modular, $\mathscr{J}(L(\Omega(L)))$ is modular by (15.5.4). Hence $L(\Omega(L))$ is modular by (14.1). Moreover, (15.10.1) holds since $L(\Omega(L))$ satisfies (ε) of (9.2), and hence (15.10.2) holds by (9.3). □

Lemma (15.11). *If an atomistic lattice L satisfies the following condition* (in particular, if L is strongly planar or satisfies (δ) of (9.2)):

(15.11.1) *If* $p, q \in \Omega(L)$ *and* $p \leqq a \vee q$ *then there exists a finite element* a_1 *such that* $p \leqq a_1 \vee q$ *and* $a_1 \leqq a$,

then the mapping $a \to \omega(a)$ *in* (15.5) *has moreover the following two properties:*

(15.11.2) *If* $p \in \Omega(L)$ *then* $\omega(a \vee p) = \omega(a) \vee \omega(p)$ *for every* $a \in L$ (hence if $b \in \mathscr{J}(L)$ then $\omega(a \vee b) = \omega(a) \vee \omega(b)$).

(15.11.3) *If* $a \lessdot b$ *in* L *then* $\omega(a) \lessdot \omega(b)$ *in* $L(\Omega(L))$.

Proof. (I) Evidently $\omega(a \vee p) \geqq \omega(a) \vee \omega(p)$. If $q \in \omega(a \vee p)$, then by (15.11.1) there exist a finite number of atoms $r_1, \ldots, r_n$ such that

$$q \leqq r_1 \vee \cdots \vee r_n \vee p \quad \text{and} \quad r_i \leqq a \ \ (i = 1, \ldots, n).$$

Hence $q \in \omega(a) \vee \omega(p)$. Therefore (15.11.2) holds.

(II) Let $a \lessdot b$. Evidently $\omega(a) < \omega(b)$. If $\omega(a) < \omega \leqq \omega(b)$ in $L(\Omega(L))$, then there exists $p \in \omega$ with $p \notin \omega(a)$. Since $a < a \vee p \leqq b$, we have $b = a \vee p$. Hence by (I)

$$\omega(b) = \omega(a) \vee \omega(p) \leqq \omega .$$

Therefore $\omega(a) \lessdot \omega(b)$. □

Theorem (15.12). *Let L be an atomistic lattice. The mapping* $a \to \omega(a)$ *of L into* $L(\Omega(L))$ *is an isomorphism if and only if L satisfies the following condition:*

(15.12.1) *If* $p \in \Omega(L)$ *and* $p \leqq a \vee b$ *then there exist two finite elements* a_1 *and* b_1 *such that* $p \leqq a_1 \vee b_1$, $a_1 \leqq a$ *and* $b_1 \leqq b$.

If L satisfies (15.12.1) *then*

(15.12.2) $(a, b) M$ *in L if and only if* $(\omega(a), \omega(b)) M$ *in* $L(\Omega(L))$.

Proof. (I) It is evident that (15.12.1) is equivalent to

$$\omega(a \vee b) = \omega(a) \vee \omega(b) \quad \text{for all } a, b \in L.$$

Hence, by (15.5.2), (15.12.1) is equivalent to that $a \to \omega(a)$ is an isomorphism.

(II) Assume that L satisfies (15.12.1). If $(\omega(a), \omega(b)) M$ in $L(\Omega(L))$, then $(a, b) M$ in L since L is isomorphic to the sublattice $\omega(L)$ of $L(\Omega(L))$. Conversely, let $(a, b) M$ in L, and let $\omega \leqq \omega(b)$ in $L(\Omega(L))$. Evidently

$$(\omega \vee \omega(a)) \wedge \omega(b) \geqq \omega \vee (\omega(a) \wedge \omega(b)).$$

If $p \in (\omega \vee \omega(a)) \wedge \omega(b)$, then there exist atoms $q_1, \ldots, q_m \in \omega$ such that $p \leqq q_1 \vee \cdots \vee q_m \vee a$. Putting $c = q_1 \vee \cdots \vee q_m$, we have $c \leqq b$, since $q_i \in \omega(b)$. Since $(a,b)M$, we have

$$p \leqq (c \vee a) \wedge b = c \vee (a \wedge b).$$

Since $a \to \omega(a)$ is an isomorphism,

$$p \in \omega(c \vee (a \wedge b)) = \omega(c) \vee (\omega(a) \wedge \omega(b)) \leqq \omega \vee (\omega(a) \wedge \omega(b)).$$

Therefore $(\omega(a), \omega(b))M$ in $L(\Omega(L))$. □

Corollary (15.13). *If an AC-lattice L satisfies* (15.12.1), *then L is M-symmetric.*

Proof. If $(a,b)M$ in L, then by (15.12.2) we have $(\omega(a), \omega(b))M$ in $L(\Omega(L))$. Since $L(\Omega(L))$ is a matroid lattice by (15.7), we have $(\omega(b), \omega(a))M$, whence $(b,a)M$ in L. □

Remark (15.14). It follows from (15.10.2) that a finite-modular AC-lattice L is modular if and only if the mapping $a \to \omega(a)$ is an isomorphism.

Theorem (15.15). *Let Λ be a given matroid lattice having the lattice operations $\sqcup$ and $\sqcap$. Assume that a subset L of Λ satisfies the following three conditions:*

(15.15.1) $\quad 0 \in L$ *and* $1 \in L$.

(15.15.2) $\quad$ *If* $a_\alpha \in L$ *for every* $\alpha \in I$ *then* $\bigsqcap(a_\alpha; \alpha \in I) \in L$.

(15.15.3) $\quad$ *If* $a \in L$ *then* $a \sqcup p \in L$ *for every atom p of* Λ.

If in L we give the same order as Λ, then L is a complete AC-lattice where the lattice operations $\vee$ and $\wedge$ satisfies the following conditions:

(15.15.4) $\quad \bigwedge_\alpha a_\alpha = \bigsqcap_\alpha a_\alpha$ *for all* $a_\alpha \in L$, *and*

(15.15.5) $$a \vee b = \begin{cases} a \sqcup b & \text{if } a \sqcup b \in L. \\ \bigsqcap(x \in L; x \geqq a, b) > a \sqcup b & \text{if } a \sqcup b \notin L, \end{cases}$$

and then Λ is isomorphic to the lattice $L(\Omega(L))$ of subspaces of $\Omega(L)$.

Moreover the following statements hold.

(15.15.6) *If Λ is weakly modular then L is strongly planar.*

(15.15.7) *If Λ is modular then L is finite-modular and then $(a,b)M^*$ in L is equivalent to $a \sqcup b \in L$.*

Proof. (I) By (15.15.2) any subset S of L has its meet in L which is equal to $\bigsqcap(a; a \in S)$ (if S is empty then $\bigsqcap(a; a \in S) = 1 \in L$ by (15.15.1)). Hence L is a complete lattice satisfying (15.15.4). Evidently (15.15.5) holds. By (15.15.1) and (15.15.3), every atom of Λ belongs to L. Hence we have $\Omega(\Lambda)=\Omega(L)$. L is atomistic since Λ is atomistic. We shall show that L has the covering property. Let $p\in\Omega(L)$ and $p\nleqq a$ in L. If $a<c \leqq a\vee p$, then since $a\vee p=a\sqcup p$ by (15.15.3) and since $a\lessdot a\sqcup p$ in Λ, we have $c=a\sqcup p\geqq p$. Hence $c=a\vee p$. Therefore L is a complete AC-lattice.

(II) For a finite number of atoms $p_1,\dots,p_n\in\Omega(\Lambda)=\Omega(L)$, by (15.15.3) we have $p_1\vee\cdots\vee p_n=p_1\sqcup\cdots\sqcup p_n$. Hence $L(\Omega(L))=L(\Omega(\Lambda))$. By (15.5) we have $L(\Omega(L))\cong\Lambda$.

(III) Let Λ be weakly modular. If $p,q,r\in\Omega(L)$ and if $p\leqq q\vee a$ and $r\leqq a$ in L, then since Λ is strongly planar by (14.5) and since $p\leqq q\sqcup a$ in Λ, there exists $s\in\Omega(\Lambda)=\Omega(L)$ such that

$$p\leqq q\sqcup r\sqcup s=q\vee r\vee s \quad \text{and} \quad s\leqq a.$$

Hence L is strongly planar.

(IV) Let Λ be modular. Since Λ satisfies (δ) of (9.2), we can show by the similar way as above that L satisfies (δ) of (9.2). Hence L is finite-modular. By (15.10.2), $(a,b)M^*$ in L is equivalent to

$$\omega(a\vee b)=\omega(a)\vee\omega(b). \tag{1}$$

Since Λ is isomorphic to $L(\Omega(L))$ by the mapping $a\to\omega(a)$, we have $\omega(a)\vee\omega(b)=\omega(a\sqcup b)$. Hence (1) is equivalent to $a\vee b=a\sqcup b$ and hence it is equivalent to $a\sqcup b\in L$ by (15.15.5). □

Exercise 15.1. Let L be an atomistic lattice. Prove that (15.11.2) implies (15.11.1) and that if L has the covering property then (15.11.3) implies (15.11.2).

Exercise 15.2. Prove that if a lattice L with 0 is weakly modular (or modular) then so is the lattice $I(L)$ of ideals of L.

Exercise 15.3. Prove that if an AC-lattice L satisfies (15.12.1) and is M^*-symmetric then L is modular.

16. Projective Spaces and Modular Matroid Lattices

Definition (16.1). A set Ω, whose elements are called *points*, is called a *projective space* if there exists a family of subsets of Ω, called *lines*, satisfying the following two conditions.

(PS 1) Every line contains at least two points, and two different points p and q are contained in one and only one line, which is denoted by $\overline{pq}$.

(PS 2) Let p, q and r be points which are not contained in one line. If s and t are different points such that $s \in \overline{pq}$ and $t \in \overline{qr}$ then there exists a point u such that $u \in \overline{pr} \cap \overline{st}$. In other words, if a line intersects two sides of a triangle at different points then it also intersects the third side.

A subset ω of a projective space Ω is called *linear* when $p, q \in \omega$ $(p \neq q)$ implies $\overline{pq} \subset \omega$. Any singleton set and the empty set are linear. By (PS 1) any line is a linear subset. The set of all linear subsets of Ω is denoted by $L(\Omega)$.

Lemma (16.2). *For two non-empty subsets ω_1 and ω_2 of a projective space Ω, we denote the set $\{p \in \Omega; p \in \overline{q_1 q_2}$ with $q_1 \in \omega_1$ and $q_2 \in \omega_2\}$ by $\omega_1 + \omega_2$.* (When $\omega_1 = \omega_2 = \{p\}$, let $\omega_1 + \omega_2 = \{p\}$.) *If ω_1, ω_2 and ω_3 are non-empty subsets of Ω, then*

(16.2.1) $$(\omega_1 + \omega_2) + \omega_3 = \omega_1 + (\omega_2 + \omega_3).$$

If ω_1 and ω_2 are linear, so is $\omega_1 + \omega_2$.

Proof. (I) Let $p \in \omega_1 + (\omega_2 + \omega_3)$, and we shall show $p \in (\omega_1 + \omega_2) + \omega_3$. There exist $q_1 \in \omega_1$ and $r \in \omega_2 + \omega_3$ with $p \in \overline{q_1 r}$, and there exist $q_2 \in \omega_2$ and $q_3 \in \omega_3$ with $r \in \overline{q_2 q_3}$ (except in the trivial case $\omega_2 = \omega_3 = \{r\}$). When $p = q_3$, then $p \in \omega_3$. When $q_1 = q_2$, then $p \in \overline{q_2 r} = \overline{q_2 q_3} \subset \omega_2 + \omega_3$. When $q_1 \neq q_2$ and $r \in \overline{q_1 q_2}$, then $p \in \overline{q_1 q_2} \subset \omega_1 + \omega_2$. Hence we may assume $p \neq q_3$, $q_1 \neq q_2$ and $r \notin \overline{q_1 q_2}$. Since q_1, q_2 and r are not contained in one line and since $p \in \overline{q_1 r}$, $q_3 \in \overline{q_2 r}$ and $p \neq q_3$, it follows from (PS 2) that there exists a point s such that $s \in \overline{q_1 q_2} \cap \overline{p q_3}$. Then $s \neq q_3$, for otherwise $r \in \overline{q_2 s} = \overline{q_1 q_2}$, a contradiction. Hence $p \in \overline{s q_3}$. Since $s \in \overline{q_1 q_2} \subset \omega_1 + \omega_2$ and $q_3 \in \omega_3$, we have $p \in (\omega_1 + \omega_2) + \omega_3$. Therefore $\omega_1 + (\omega_2 + \omega_3) \subset (\omega_1 + \omega_2) + \omega_3$, and by symmetry of ω_1 and ω_3 we obtain (16.2.1).

(II) Let ω_i $(i = 1, 2)$ be linear. Then $\omega_i + \omega_i = \omega_i$. If $p, q \in \omega_1 + \omega_2$ $(p \neq q)$, then by (I)

$$\begin{aligned}\overline{pq} &\subset (\omega_1 + \omega_2) + (\omega_1 + \omega_2) = \{(\omega_1 + \omega_2) + \omega_1\} + \omega_2 \\ &= \{(\omega_1 + \omega_1) + \omega_2\} + \omega_2 = \omega_1 + (\omega_2 + \omega_2) = \omega_1 + \omega_2.\end{aligned}$$

Hence $\omega_1 + \omega_2$ is linear. □

Theorem (16.3). *If Ω is a projective space, then the set $L(\Omega)$ of all linear subsets of Ω forms a modular matroid lattice, ordered by set-inclusion, where $\bigwedge(\omega_\alpha; \alpha \in I)$ is the intersection of linear sets ω_α and for two different non-empty linear subsets ω_1 and ω_2*

$$\omega_1 \vee \omega_2 = \{p \in \Omega; p \in \overline{q_1 q_2} \text{ with } q_1 \in \omega_1 \text{ and } q_2 \in \omega_2\}.$$

Proof. (I) It is easy to verify that $L(\Omega)$ satisfies the three conditions in (15.3). Hence $L(\Omega)$ forms an upper continuous lattice. $L(\Omega)$ is atomistic since any singleton set of Ω is an atom of $L(\Omega)$. Thus $L(\Omega)$ is a compactly atomistic lattice.

(II) For any two different non-zero elements ω_1 and ω_2 of $L(\Omega)$, it follows from (16.2) that

$$\omega_1 \vee \omega_2 = \omega_1 + \omega_2 .$$

Let $\omega_3 \leqq \omega_2$ and let $p \in (\omega_3 \vee \omega_1) \wedge \omega_2$. There exist $q_3 \in \omega_3$ and $q_1 \in \omega_1$ with $p \in \overline{q_3 q_1}$ (except in the trivial case). When $p = q_3$, then $p \in \omega_3 \leqq \omega_3 \vee (\omega_1 \wedge \omega_2)$. When $p \neq q_3$, since

$$q_1 \in \overline{q_3 p} \subset \omega_3 \vee \omega_2 = \omega_2 ,$$

we have $q_1 \in \omega_1 \wedge \omega_2$. Hence $p \in \overline{q_3 q_1} \subset \omega_3 \vee (\omega_1 \wedge \omega_2)$. Therefore $(\omega_3 \vee \omega_1) \wedge \omega_2 \leqq \omega_3 \vee (\omega_1 \wedge \omega_2)$. Thus $(\omega_1, \omega_2) M$ in $L(\Omega)$, and $L(\Omega)$ is modular. □

Lemma (16.4). *Let L be a finite-modular AC-lattice. The atom space $\Omega(L)$ of L forms a projective space where*

$$\overline{pq} = \{ r \in \Omega(L); r \leqq p \vee q \} .$$

A subset ω of $\Omega(L)$ is linear if and only if ω is a subspace of $\Omega(L)$.

Proof. (I) In the atom space $\Omega(L)$, let the set $\overline{pq} = \{ r \in \Omega(L); r \leqq p \vee q \}$ be a line for every pair of different points p and q. We shall show that (PS 1) holds. Evidently $\overline{pq}$ contains at least two points. Let two different points p and q be contained in a line $\overline{rs}$. When $r = p$, we have $q \leqq p \vee s$. Hence by the exchange property $p \vee s = p \vee q$, which implies $\overline{pq} = \overline{rs}$. When $r \neq p$, by the exchange property $p \leqq r \vee s$ implies $r \vee p = r \vee s$. Hence $q \leqq r \vee p$ and by the exchange property again we get $p \vee q = r \vee p$. Hence $\overline{pq} = \overline{rs}$. Thus (PS 1) holds.

(II) We shall show that (PS 2) holds. Let $s \in \overline{pq}$, $t \in \overline{qr}$ and $s \neq t$. When $s = p$, then $u = p$ may be used. When $s \neq p$, by the exchange property we have $q \leqq p \vee s$, whence

$$t \leqq q \vee r \leqq s \vee p \vee r .$$

It follows from (δ) of (9.2) that there exists a point u such that

$$t \leqq s \vee u \quad \text{and} \quad u \leqq p \vee r .$$

Then $u \in \overline{pr}$, and by the exchange property $u \in \overline{st}$. Thus (PS 2) holds.

(III) A subset ω is linear if and only if ω satisfies (15.2.1), which is equivalent to (15.1.1). □

We remark that this theorem is valid when L is an atomistic lattice which has the finite covering property and satisfies one of the statements in (9.7).

Theorem (16.5). *If L is a modular matroid lattice, then the atom space $\Omega(L)$ forms a projective space, and L is isomorphic to the lattice $L(\Omega(L))$ of linear subsets of $\Omega(L)$.*

Proof. By (16.4), $\Omega(L)$ forms a projective space and $L(\Omega(L))$ coincides with the lattice of subspaces of $\Omega(L)$. Since L is compactly atomistic, L is isomorphic to $L(\Omega(L))$ by (15.5). ▯

Theorem (16.6). *Let L be a modular matroid lattice.*

(16.6.1) *L is the direct sum of irreducible sublattices $\{L[0,z_\alpha];\alpha\in I\}$ where $z_\alpha\in Z(L)$ for every $\alpha\in I$.*

(16.6.2) *Two different atoms p and q of L are contained in the same summand $L[0,z_\alpha]$ if and only if $p\vee q$ contains a third atom.*

(16.6.3) *L is irreducible if and only if for any different atoms p and q of L, $p\vee q$ contains a third atom.*

Proof. This follows from (13.6) and (11.6). ▯

Definition (16.7). A projective space Ω is called *irreducible* when every line of Ω contains at least three points.

It follows from (16.3), (16.5), and (16.6.3) that a projective space Ω is irreducible if and only if the lattice $L(\Omega)$ of linear subsets of Ω is irreducible and that a modular matroid lattice L is irreducible if and only if the projective space $\Omega(L)$ of atoms of L is irreducible.

Remark (16.8). The lattice $L(\Omega)$ of linear subsets of a projective space is called a *generalized projective geometry* on Ω; when Ω is irreducible, $L(\Omega)$ is called a *projective geometry* on Ω. By (16.5), a modular matroid lattice L is isomorphic to the generalized projective geometry on $\Omega(L)$. L is sometimes called a *projective lattice*.

Theorem (16.9). *Let E be a vector space over a division ring K. Then the set L(E) of all subspaces of E forms an irreducible modular matroid lattice, ordered by set-inclusion, where the meet of subspaces is their intersection and the join of subspaces A_1 and A_2 is the linear sum*

$$A_1+A_2=\{x+y;x\in A_1 \text{ and } y\in A_2\}.$$

Proof. (I) It is easy to verify that $L(E)$ satisfies the three conditions in (15.3). Hence $L(E)$ forms an upper continuous lattice. $L(E)$ is atomistic since any one-dimensional subspace of E is an atom of $L(E)$.

(II) For any subspaces A_1 and A_2, since A_1+A_2 is a subspace, it is the join of A_1 and A_2 in $L(E)$. Let $A_3\leqq A_2$, and let $x\in(A_3\vee A_1)\wedge A_2$. Since $x\in A_3+A_1$, there exist $y_3\in A_3$ and $y_1\in A_1$ such that $x=y_3+y_1$. Then $y_1=x-y_3\in A_2+A_3=A_2$, whence $y_1\in A_1\wedge A_2$. Hence $x\in A_3\vee(A_1\wedge A_2)$. Therefore $L(E)$ is modular.

(III) $L(E)$ is irreducible since the join of two different atoms Kx and Ky contains a third atom $K(x+y)$. ▯

Corollary (16.10). *Let E be a vector space. Then the set Ω of atoms of $L(E)$ forms an irreducible projective space and $L(E)$ is isomorphic to the projective geometry $L(\Omega)$.*

Proof. This is a consequence of (16.9) and (16.5). ▯

EXERCISE 16.1. Show that any projective space is a union of disjoint irreducible projective subspaces.

References for Chapter III

For Section 13: M. L. Dubreil-Jacotin, L. Leisieur and R. Croisot [1], S. MacLane [1], F. Maeda [7], U. Sasaki and S. Fujiwara [1].

For Section 14: U. Sasaki [2], B. Jónsson [1], F. Maeda [7].

For Section 15: M. F. Janowitz [5], R. Wille [1].

For Section 16: F. Maeda [2] (Chapter III), R. Baer [1].

Chapter IV

Parallelism in Symmetric Lattices

In this chapter we use the term "point" instead of "atom", for the arguments here are geometric. Moreover, an element a of an AC-lattice is called a line (resp. a plane) when $h(a)=2$ (resp. $h(a)=3$), as in a matroid lattice.

17. Parallelism in Lattices

Definition (17.1). Let L be a lattice with 0 and let a and b be non-zero elements of L. We write $a<|b$ when

(17.1.1) $$a \wedge b=0 \quad \text{and} \quad b \lessdot a \vee b.$$

When $a<|b$ and $b<|a$, we say that a and b are *parallel* and write $a \| b$.

Remark (17.2). In a lattice L with the covering property, if p is a point and if $p \wedge a=0$, then by the covering property we have $a \lessdot a \vee p$. Hence $p<|a$. In particular, if p and q are different points of L then $p \| q$.

Remark (17.3). Let a and b be non-zero elements of an AC-lattice L. If $a<|b$ and if b is finite, then a is finite and $h(a) \leqq h(b)$. This is so because, since $a<a \vee b$ and $b \lessdot a \vee b$, we have $h(a)<h(a \vee b)=h(b)+1$, whence $h(a) \leqq h(b)$. Moreover, two finite elements a and b are parallel if and only if $a \wedge b=0$ and $h(a)=h(b)=h(a \vee b)-1$.

If two lines l and k are contained in a plane and if $l \wedge k=0$, then evidently they are parallel.

Lemma (17.4). *In a lattice L with 0, if $a<|b$ then $a_1 \vee b=a \vee b$ for any $a_1 \in L$ with $0<a_1 \leqq a$.*

Proof. If $a<|b$ and $0<a_1 \leqq a$, then since $a_1 \wedge b \leqq a \wedge b=0$, we have $a_1 \nleqq b$. Hence $b<a_1 \vee b \leqq a \vee b$. Since $b \lessdot a \vee b$, we have $a_1 \vee b =a \vee b$. □

Remark (17.5). If $a<|b$ in a lattice with 0, then for any pair of non-zero elements a_1 and a_2 contained in a, we have $a_1 \sim_b a_2$ by (17.4).

Lemma (17.6). *Let $a<|b$ in a lattice L with 0.*

(17.6.1) *If $0<a_1 \leqq a$ and if a_1 is not a point then $(b, a_1)M$ does not hold.*

(17.6.2) *If $a_1 \leqq a$ then $(a_1, b)M^*$ holds.*

(17.6.3) *If $0<a_1<a$ then $(b, a_1)M^*$ does not hold.*

Proof. (I) If $a_1>0$ is not a point, then there exists $c\in L$ such that $0<c<a_1$. If moreover $a_1\leqq a$, then $c\vee b=a_1\vee b=a\vee b$ by (17.4) and $b\wedge a_1=0$. Hence

$$(c\wedge b)\wedge a_1=(a_1\vee b)\wedge a_1=a_1>c=c\vee(b\wedge a_1).$$

Thus $(b,a_1)M$ does not hold.

(II) If $a_1=0$ then evidently $(a_1,b)M^*$. If $0<a_1\leqq a$, then $b\lessdot a\wedge b=a_1\vee b$ by (17.4). Hence $(a_1,b)M^*$ by (7.5.2).

(III) If $0<a_1<a$, then

$$a\wedge(b\vee a_1)=a\wedge(b\vee a)=a>a_1=(a\wedge b)\vee a_1.$$

Hence $(b,a_1)M^*$ does not hold. □

Lemma (17.7). *If a lattice L with 0 is M^*-symmetric then $a<|b$ implies that a is a point in L.*

Proof. If $a<|b$, then since L is M^*-symmetric, it follows from (17.6.2) that $(b,a_1)M^*$ holds for every $a_1\leqq a$. Hence a must be a point by (17.6.3). □

Lemma (17.8). *Let l be a line of an AC-lattice L. Then, $l<|a$ if and only if $(a,l)\bar{M}$.*

Proof. If $l<|a$ then $(a,l)\bar{M}$ by (17.6.1). Conversely. if $(a,l)\bar{M}$, then there exists an element $c\leqq l$ such that

$$c\vee(a\wedge l)<(c\vee a)\wedge l.$$

Evidently $0<c<l$, whence $h(c)=1$. Since $c\leqq c\vee(a\wedge l)<(c\vee a)\wedge l\leqq l$, we have

$$c=c\vee(a\wedge l)\quad\text{and}\quad(c\vee a)\wedge l=l,\quad\text{whence } a\wedge l\leqq c\quad\text{and}\quad l\leqq c\vee a.$$

If we had $a\wedge l\neq 0$, then $a\wedge l=c$ since $h(c)=1$, and then $l\leqq(a\wedge l)\vee a=a$, whence $l=a\wedge l=c$, a contradiction. Hence $a\wedge l=0$. Moreover we have $l\vee a=c\vee a\gtrdot a$ by the covering property. Therefore $l<|a$. □

Theorem (17.9). *Let L be an AC-lattice. The following four statements are equivalent.*

(α) *L is finite-modular.*
(β) *L is M^*-symmetric.*
(γ) *$a<|b$ implies that a is a point in L* (in other words, L has only trivial parallelism).
(δ) *There exists no line l such that $l<|a$ for some $a\in L$.*

Proof. The equivalence of (α) and (β) follows from (9.5). The implication (β)⇒(γ) follows from (17.7), and (γ)⇒(δ) is obvious. If (δ) holds, then it follows from (17.8) that L satisfies the statement (β) of (9.2). Hence L is finite-modular. $\square$

Corollary (17.10). *A matroid lattice L is modular if and only if there exists no line l such that $l <\!| a$ for some $a \in L$.*

Proof. This follows from (17.9) and (14.1). $\square$

Lemma (17.11). *In a lattice L with 0, if $a <\!| b$ and $0 < a_1 \leqq a$ then $a_1 <\!| b$.*

Proof. Let $a <\!| b$ and $0 < a_1 \leqq a$. We have $a_1 \wedge b \leqq a \wedge b = 0$. By (17.4) we have $a_1 \vee b = a \vee b > b$. Hence $a_1 <\!| b$. $\square$

Lemma (17.12). *In an atomic lattice L with the covering property, if $a <\!| b$ and $b \leqq b_2$ then either $a <\!| b_2$ or $a < b_2$.*

Proof. Let $a <\!| b$ and $b \leqq b_2$. Since L is atomic there is a point p with $p \leqq a$. When $a \wedge b_2 = 0$, we have $p \wedge b_2 = 0$. By (17.4) we have $p \vee b = a \vee b$, whence $p \vee b_2 = a \vee b_2$. By the covering property, $b_2 \lessdot p \vee b_2 = a \vee b_2$. Hence $a <\!| b_2$.

When $a \wedge b_2 \neq 0$, by (17.4) we have

$$a < a \vee b = (a \wedge b_2) \vee b \leqq b_2 . \quad \square$$

Lemma (17.13). *Let a and b be non-zero elements of an atomic lattice L with the covering property.*

(17.13.1) *$a <\!| b$ if and only if $a \wedge b = 0$ and there exists a point p such that $p \leqq a$ and $p \vee b = a \vee b$.*

(17.13.2) *$a \parallel b$ if and only if $a \wedge b = 0$ and there exist points p and q such that $p \leqq a$, $q \leqq b$ and $a \vee q = b \vee p$.*

Proof. (I) Let $a <\!| b$. Since L is atomic there is a point p with $p \leqq a$. By (17.4) we have $p \vee b = a \vee b$. Conversely, if $a \wedge b = 0$ and p is a point such that $p \leqq a$ and $p \vee b = a \vee b$, then since $p \wedge b \leqq a \wedge b = 0$, by the covering property we have $b \lessdot p \vee b = a \vee b$. Hence $a <\!| b$.

(II) If $a \parallel b$ then by (I) there exist points $p \leqq a$ and $q \leqq b$ such that $a \vee q = a \vee b = b \vee p$. Conversely, if $a \wedge b = 0$, $p \leqq a$, $q \leqq b$ and $a \vee q = b \vee p$, then since $b \leqq a \vee q$, we have $a \vee b \leqq a \vee q \leqq a \vee b$. Hence $a \vee q = a \vee b$, and similarly $b \vee p = a \vee b$. Therefore we have $a <\!| b$ and $b <\!| a$ by (I). $\square$

Lemma (17.14). *In a weakly modular lattice L with the covering property, if $a <\!| b$ and if q is a point with $q \leqq b$, then $a \parallel (a \vee q) \wedge b$.*

Proof. Put $b_1 = (a \vee q) \wedge b$. Then $a \wedge b_1 \leqq a \wedge b = 0$. Since $a \vee b_1 \leqq a \vee q$ and $q \leqq b_1$, we have $a \vee b_1 = a \vee q$. To prove $a \parallel b_1$, it suffices to show $a \lessdot a \vee q$ and $b_1 \lessdot a \vee q$. Since $a \wedge q \leqq a \wedge b = 0$, we have $a \lessdot a \vee q$ by the covering property.

Next, take $c \in L$ with $b_1 \leqq c \leqq a \vee q$. When $c \leqq b$, we have $c \leqq (a \vee q) \wedge b = b_1$, whence $c = b_1$. When $c \nleqq b$, since

$$b < b \vee c \leqq b \vee a \vee q = a \vee b \quad \text{and} \quad b \lessdot a \vee b,$$

we have $b \vee c = a \vee b$, whence $a \vee q \leqq b \vee c$. Since L is weakly modular and $(a \vee q) \wedge b \geqq q > 0$, we have $(b, a \vee q)M$. Hence

$$c = c \vee b_1 = c \vee \{b \wedge (a \vee q)\} = (c \vee b) \wedge (a \vee q) = a \vee q.$$

Therefore $b_1 \lessdot a \vee q$. □

Theorem (17.15). *In a weakly modular lattice L with the covering property, assume that a and b are parallel and that p and q are points with $p \leqq a$ and $q \leqq b$. Put*

$$\varphi(a_1) = (a_1 \vee q) \wedge b \quad \textit{for } a_1 \in L[p, a], \quad \textit{and}$$
$$\psi(b_1) = (b_1 \vee p) \wedge a \quad \textit{for } b_1 \in L[q, b].$$

Then φ and ψ are mutually inverse, isomorphic mappings between the lattices $L[p, a]$ and $L[q, b]$.

Two elements $a_1 \in L[p, a]$ and $b_1 \in L[q, b]$ correspond by these mappings if and only if the following equality holds:

$$(1) \qquad a_1 \vee q = b_1 \vee p.$$

In this case $a_1 \parallel b_1$.

Proof. (I) It is evident that $\varphi(a_1) \in L[q, b]$ and $\psi(b_1) \in L[p, a]$. Since by (17.11) we have $a_1 <| b$, it follows from (17.14) that $a_1 \parallel \varphi(a_1)$. Similarly we have $b_1 \parallel \psi(b_1)$. Since $a_1 \parallel \varphi(a_1)$, $p \leqq a_1$ and $q \leqq \varphi(a_1)$, by (17.4) we have $p \vee \varphi(a_1) = a_1 \vee \varphi(a_1) = a_1 \vee q$. Similarly we have $q \vee \psi(b_1) = b_1 \vee p$. Thus (1) holds when either $\varphi(a_1) = b_1$ or $\psi(b_1) = a_1$.

(II) Conversely, assume that (1) holds. Since $(p, b)M$ by (7.6), we have

$$\varphi(a_1) = (a_1 \vee q) \wedge b = (b_1 \vee p) \wedge b = b_1 \vee (p \wedge b) = b_1.$$

Similarly we have $\psi(b_1) = a_1$.

(III) For any $a_1 \in L[p, a]$, if we put $b_1 = \varphi(a_1)$, then since (1) holds by (I), we have $\psi(b_1) = a_1$ by (II). Hence $\psi\varphi(a_1) = a_1$. Similarly $\varphi\psi(b_1) = b_1$ for any $b_1 \in L[q, b]$. Therefore φ and ψ are mutually inverse, one-to-one mappings between $L[p, a]$ and $L[q, b]$. Since they preserve the order, they are isomorphic mappings. □

Definition (17.16). We call φ and ψ in (17.15) *parallel mappings* between $L[p, a]$ and $L[q, b]$.

Theorem (17.17). *Let L be a matroid lattice. The following three statements are equivalent.*

(α) *L is weakly modular.*
(β) *If $a <\mid b$ in L and if q is a point of L with $q \leqq b$ then $a \parallel (a \vee q) \wedge b$.*
(γ) *If $a <\mid b$ in L and if q is a point of L with $q \leqq b$ then there exists $b_1 \in L$ such that $a \parallel b_1$ and $q \leqq b_1 \leqq b$.*

Proof. The implication (α)⇒(β) follows from (17.14). (β)⇒(γ) is trivial.

(γ)⇒(α). Assume that (γ) holds. In order to prove (α), by (14.5) it suffices to prove that

(1) if p, q and r are points such that $p \leqq q \vee a$ and $r \leqq a$ then there exists a point s such that

$$p \leqq q \vee r \vee s \quad \text{and} \quad s \leqq a.$$

We may assume that $p \neq q$ and $q \nleqq a$, for otherwise (1) holds evidently. Let $l = p \vee q$. When $l \wedge a \neq 0$, we take a point s with $s \leqq l \wedge a$. Since $s \leqq p \vee q$ and $s \wedge q \leqq a \wedge q = 0$, by the exchange property we have $p \leqq q \vee s$. Thus (1) holds. When $l \wedge a = 0$, since by the covering property

$$a \lessdot a \vee q = a \vee p \vee q = a \vee l,$$

we have $l <\mid a$. By (γ) there exists $k \in L$ such that

$$l \parallel k \quad \text{and} \quad r \leqq k \leqq a.$$

Since by (17.3) k is a line, there exists a point s such that $k = r \vee s$. By (17.4) we have $p \vee k = q \vee k = l \vee k$. Hence $p \leqq q \vee k = q \vee r \vee s$. Thus (1) holds. □

Corollary (17.18). *Let L be a matroid lattice. The following two statements are equivalent.*

(α) *In the irreducible decomposition $L = \dot{\bigcup}(L[0, z_\alpha]; \alpha \in I)$, all summands $L[0, z_\alpha]$ are weakly modular.*
(β) *If $a <\mid b$ in L and if q is a point of L with $q \leqq e(a) \wedge b$, then there exists $b_1 \in L$ such that*

$$a \parallel b_1 \quad \text{and} \quad q \leqq b_1 \leqq b.$$

Proof. (α)⇒(β). Let $a <\mid b$ and $q \leqq e(a) \wedge b$. Since any two points contained in a are perspective by (17.5), it follows from (13.6.2) that a is contained in an irreducible sublattice $L[0, z_\alpha]$, whence $e(a) = z_\alpha$. Putting $b' = e(a) \wedge b$, we have $b' \lessdot a \vee b'$, because if $b' \leqq c \leqq a \vee b'$ then since $b \leqq b \vee c \leqq a \vee b$ and $b \lessdot a \vee b$ we have $b \vee c = b$ or $a \vee b$, whence

$c=b'$ or $a \vee b'$. Hence in a weakly modular lattice $L[0, z_\alpha]$ we have $a <| b'$ and $q \leqq b'$. By (17.17) there exists b_1 such that $a \| b_1$ and $q \leqq b_1 \leqq b' \leqq b$.

$(\beta) \Rightarrow (\alpha)$. If $a <| b$ in an irreducible sublattice $L[0, z_\alpha]$, then $e(a) = z_\alpha$ as above. Hence if L satisfies (β) then $L[0, z_\alpha]$ satisfies (γ) in (17.17). Hence $L[0, z_\alpha]$ is weakly modular. □

Theorem (17.19). *A matroid lattice L is weakly modular if and only if L satisfies the following condition:*

(17.19.1) *If p and q are different points such that $p \sim_x q$ and if r is a point with $r \leqq x$, then either the line $p \vee q$ contains a third point s with $s \leqq x$, or there exists a line l such that $p \vee q \| l$ and $r < l \leqq x$.*

Proof. By (14.6) it suffices to show that (14.6.2) and (17.19.1) are equivalent.

(14.6.2)$\Rightarrow$(17.19.1). If $p \neq q$, $p \sim_x q$ and $r \leqq x$, then by (14.6.2) there exists either a point s such that $p \sim_s q$ and $s \leqq x$ or a line l such that $p \sim_l q$ and $r < l \leqq x$. In the former case, $p \vee q$ contains a third point $s \leqq x$. In the latter case, since $p \leqq q \vee l$, the two lines $p \vee q$ and l are contained in the plane $q \vee l$. When $(p \vee q) \wedge l \neq 0$, taking a point $s \leqq (p \vee q) \wedge l$, we have $p \neq s$ and $q \neq s$ since $p \wedge l = q \wedge l = 0$. Hence $p \vee q$ contains a third point s and we have $s < l \leqq x$. When $(p \vee q) \wedge l = 0$, by (17.3) we have $p \vee q \| l$. Thus (17.19.1) holds.

(17.19.1)$\Rightarrow$(14.6.2). If $p \neq q$, $p \sim_x q$ and $r \leqq x$, then by (17.19.1), either the line $p \vee q$ contains a third point s with $s \leqq x$ or there exists a line l such that $p \vee q \| l$ and $r < l \leqq x$. In the former case, we have $p \sim_s q$ by the covering property. In the latter case, we have $p \sim_l q$ by (17.5). □

Exercise 17.1. (Parallelism and distributivity). Let $a <| b$ in a lattice with 0 and assume that a is not a point. Prove the following statements.

(1) $b \triangledown a$ does not hold.

(2) If z is neutral and $a \leqq z$, then $z \wedge b \neq 0$.

(3) If z is neutral and $b \leqq z$, then $a < z$.

Exercise 17.2. Prove that a weakly modular matroid lattice L is modular if and only if no two lines in L are parallel.

Exercise 17.3. Let L be a weakly modular lattice with the covering property. Prove that if $a \wedge b = 0$ in L and if p and q are points such that $p \leqq a$ and $q \leqq b$ then $(b \vee p) \wedge a \| (a \vee q) \wedge b$.

18. Incomplete Elements in Affine Matroid Lattices

Axiom (18.1). *Euclid's strong parallel axiom.* Let l be a line in a matroid lattice. If p is a point with $p \nleq l$, then there exists one and only one line k such that $l \parallel k$ and $p < k$.

Axiom (18.2). *Euclid's weak parallel axiom.* Let l be a line in a matroid lattice. If p is a point with $p \nleq l$, then there exists at most one line k such that $l \parallel k$ and $p < k$.

Definition (18.3). Let L be a weakly modular matroid lattice of length $\geqq 4$ (L may be of infinite length). When L satisfies the Euclid's weak parallel axiom, we call L an *affine matroid lattice.*

By (14.7), if an affine matroid lattice is not modular then it is irreducible.

Lemma (18.4). *Let L be an affine matroid lattice.*

(18.4.1) *If $a <\!| b$ in L then for any point q with $q \leqq b$ there exists one and only one element b_1 such that*

$$a \parallel b_1 \quad \textit{and } q \leqq b_1 .$$

Then $b_1 = (a \vee q) \wedge b \leqq b.$

(18.4.2) *If* $a \parallel b$, $a \parallel b'$ *and* $b \wedge b' \neq 0$ *in L, then* $b = b'$.

Proof. (I) Let $a <\!| b$ and $q \leqq b$. Putting $b_1 = (a \vee q) \wedge b$, by (17.14) we have $a \parallel b_1$ and $q \leqq b_1$.

Next, let b' be an element such that $a \parallel b'$ and $q \leqq b'$. We shall show that

(1) if r is a point with $r \neq q$ and $r \leqq b_1$ then $r \leqq b'$.

Put $k = r \vee q$. Since $b_1 \parallel a$ and k is a line with $k \leqq b_1$, taking a point p with $p \leqq a$, it follows from (17.15) that $l = (k \vee p) \wedge a$ is a line with $l \parallel k$. Moreover, since $a \parallel b'$, $l \leqq a$ and $q \leqq b'$, the element $k' = (l \vee q) \wedge b'$ is a line with $k' \parallel l$. Then, since $l \parallel k$, $l \parallel k'$, $q < k$ and $q < k'$, it follows from Euclid's weak parallel axiom that $k = k'$. Hence $r \leqq k' \leqq b'$. Thus (1) has been proved. Since L is atomistic and $q \leqq b'$, (1) implies $b_1 \leqq b'$. By symmetry we have $b_1 = b'$. This completes the proof of (18.4.1).

(II) Let $a \parallel b$, $a \parallel b'$ and $b \wedge b' \neq 0$. Taking a point q with $q \leqq b \wedge b'$, it follows from (18.4.1) that $b = b'$. ▯

Definition (18.5). In a weakly modular AC-lattice, a line l is called *complete* when there exists no line parallel to l, and l is called *incomplete* when there exists a line parallel to l. An element a of height $\geqq 2$ (a may be not finite) is called *incomplete* when every line contained in a is incomplete.

Remark (18.6). Let L be a weakly modular AC-lattice and let l be a line of L. It follows from (17.14) and (17.8) that the following three statements are equivalent.

(α) l is incomplete.
(β) $l <\!| a$ for some $a \in L$.
(γ) $(a, l)\bar{M}$ for some $a \in L$.

Hence, l is complete if and only if l is a modular element of L, and an element a of height $\geqq 2$ is incomplete if and only if it contains no modular line.

Moreover, by (17.10), a weakly modular matroid lattice L is modular if and only if every line of L is complete.

Lemma (18.7). *If a weakly modular matroid lattice L is not modular then any complete line l of L contains at least three points.*

Proof. Let p and q be different points with $p \vee q = l$. By (14.7) we have $p \sim q$. Since there is no line parallel to $p \vee q$, it follows from (17.19.1) that $p \vee q$ contains a third point. □

Lemma (18.8). *If $a <\!| b$ in a weakly modular AC-lattice, then a is either a point or an incomplete element.*

Proof. If a contains a line l, then since $l <\!| b$ by (17.11), it follows from (17.14) that there is a line parallel to l. Hence l is incomplete. □

Lemma (18.9). *In an affine matroid lattice L, if l is an incomplete line, then for any point p with $p \nleqq l$ there exists one and only one line k such that $l \parallel k$ and $p < k$.*

Proof. (I) Let l be an incomplete line. First we shall show that if d is a line parallel to l and if p is a point with $p \nleqq l \vee d$ then there exists a line k such that $l \parallel k$ and $p < k$. Since $l \parallel d$ and $d \leqq d \vee p$, it follows from (17.12) that we have either $l <\!| d \vee p$ or $l < d \vee p$. If $l < d \vee p$, then $d < d \vee l \leqq d \vee p$, whence by the covering property we would have $d \vee l = d \vee p \geqq p$, a contradiction. Hence we have $l <\!| d \vee p$. By (17.14), $k = (l \vee p) \wedge (d \vee p)$ is a line parallel to l.

(II) Let p be a point with $p \nleqq l$. Since l is incomplete, there is a line d parallel to l. When $p \nleqq l \vee d$, by (I) there exists a line k such that $l \parallel k$ and $p < k$. When $p \leqq l \vee d$, since $l \vee d$ is a plane and since L is of length $\geqq 4$, there exists a point p' such that $p' \nleqq l \vee d$. By (I) there exists a line k' such that $l \parallel k'$ and $p' < k'$. If $p \leqq l \vee k'$, then since $l \vee p' = l \vee k'$ by (17.4), we would have $p \leqq l \vee p'$, whence by the exchange property $p' \leqq l \vee p \leqq l \vee d$, a contradiction. Hence $p \nleqq l \vee k'$. By (I) there exists a line k such that $l \parallel k$ and $p < k$.

(III) The uniqueness of k follows from Euclid's weak parallel axiom. □

Remark (18.10). Let L be an affine matroid lattice. It follows from (18.9) that L satisfies Euclid's strong parallel axiom if and only if every line of L is incomplete.

Lemma (18.11). *In an affine matroid lattice L, let a be an element of height $\geqq 2$. If a has a base $\{p, p_\alpha; \alpha \in I\}$ such that the line $l_\alpha = p \vee p_\alpha$ is incomplete for every $\alpha \in I$, then for any point q with $q \not\leqq a$ there exists an element b such that $a \parallel b$ and $q < b$.*

Proof. For every $\alpha \in I$, since l_α is incomplete and $q \not\leqq l_\alpha$, by (18.9) there exists a line k_α such that $l_\alpha \parallel k_\alpha$ and $q < k_\alpha$. Putting $b = \bigvee(k_\alpha; \alpha \in I)$, we shall show $a \parallel b$. By (17.4) we have

(1) $$l_\alpha \vee q = l_\alpha \vee k_\alpha = p \vee k_\alpha \quad \text{for every } \alpha \in I.$$

Hence $a \vee q = \bigvee(l_\alpha; \alpha \in I) \vee q = \bigvee(k_\alpha; \alpha \in I) \vee p = b \vee p$. Therefore, by (17.13.2), we need only verify $a \wedge b = 0$.

Since L is upper continuous, it suffices to prove that

(2) $$\bigvee(l_\alpha; \alpha \in J) \wedge \bigvee(k_\alpha; \alpha \in J) = 0$$

for every finite subset J of I. We shall prove it by mathematical induction on the cardinal number n of J. When $n = 1$, (2) holds evidently. Assume that (2) holds when $n = m$. When $n = m + 1$, for brevity let $J = \{1, \dots, m+1\}$. We put

$$\begin{aligned} a_0 &= l_1 \vee \cdots \vee l_{m+1}, & b_0 &= k_1 \vee \cdots \vee k_{m+1}, \\ a' &= l_1 \vee \cdots \vee l_m, & b' &= k_1 \vee \cdots \vee k_m, \\ a'' &= l_1 \vee \cdots \vee l_{m-1} \vee l_{m+1}, & b'' &= k_1 \vee \cdots \vee k_{m-1} \vee k_{m+1}. \end{aligned}$$

By (1) we have $a' \vee q = b' \vee p$, and since $a' \wedge b' = 0$ by the assumption we have $b' \parallel a'$ by (17.13.2). Since $a' < a_0$, by (17.12) we have either $b' <\!\!| a_0$ or $b' < a_0$. But $b' < a_0$ is false since $q \leqq b'$ and $q \not\leqq a_0$. Hence we have $b' <\!\!| a_0$. Similarly we have $b'' \parallel a''$ and $b'' <\!\!| a_0$.

Now assume that there is a point r such that $r \leqq a_0 \wedge b_0$. By (17.14) we have $b' \parallel (b' \vee r) \wedge a_0$. Moreover $b' \vee r = b_0$, because, taking a point r_{m+1} such that $k_{m+1} = q \vee r_{m+1}$, we have

$$r \leqq b_0 = b' \vee k_{m+1} = b' \vee q \vee r_{m+1} = b' \vee r_{m+1},$$

whence by the exchange property we have $b' \vee r = b' \vee r_{m+1} = b_0$. Thus we have $a_0 \wedge b_0 \parallel b'$. Similarly we have $a_0 \wedge b_0 \parallel b''$. Since $b' \wedge b'' \geqq q > 0$, it follows from (18.4.2) that $b' = b''$. Hence

(3) $$k_{m+1} \leqq k_1 \vee \cdots \vee k_m.$$

Let $k_i = q \vee r_i$ $(i=1, \dots, m)$ and put $c = k_1 \vee \cdots \vee k_m \vee p$. Since $c = r_1 \vee \cdots \vee r_m \vee p \vee q$, we have $h(c) \leqq m+2$. On the other hand, by (3) and (1)

$$c = k_1 \vee \cdots \vee k_{m+1} \vee p = l_1 \vee \cdots \vee l_{m+1} \vee q = p_1 \vee \cdots \vee p_{m+1} \vee p \vee q.$$

Since $(p_1, \dots, p_{m+1}, p) \perp$ and $(p_1 \vee \cdots \vee p_{m+1} \vee p) \wedge q \leqq a \wedge q = 0$, we have $h(c) = m+3$, a contradiction. Therefore $a_0 \wedge b_0 = 0$. Thus (2) holds when $n = m+1$. □

Theorem (18.12). *In an affine matroid lattice L, if a is an incomplete element, then for any point q with $q \nleqq a$ there exists one and only one element b such that $a \parallel b$ and $q < b$.*

Proof. By (8.16) a has a base $\{p, p_\alpha; \alpha \in I\}$. Since a is incomplete, the line $p \vee p_\alpha$ is incomplete for every $\alpha \in I$. Hence by (18.11) there exists b such that $a \parallel b$ and $q < b$. The uniqueness of b follows from (18.4.2).

Corollary (18.13). *In an affine matroid lattice, let a be an element of height $\geqq 2$. The following three statements are equivalent.*

(α) *a is incomplete and $a \neq 1$.*
(β) *There exists an element b with $a \parallel b$.*
(γ) *There exists an element b with $a <\!| b$.*

Proof. (α)⇒(β). Since $a \neq 1$, there exists a point q with $q \nleqq a$. Hence by (18.12) there exists b with $a \parallel b$. The implication (β)⇒(γ) is trivial. (γ)⇒(α) follows from (18.8). □

Corollary (18.14). *In an affine matroid lattice L, if $a <\!| b$ and if b is an incomplete element, then there exists one and only one element a_2 such that $a_2 \parallel b$ and $a \leqq a_2$.*

Proof. Take points p and q with $p \leqq a$ and $q \leqq b$. Since b is incomplete and $p \wedge b \leqq a \wedge b = 0$, by (18.12) there exists one and only one element a_2 such that $b \parallel a_2$ and $p \leqq a_2$. It suffices to show that $a \leqq a_2$. By (17.14), putting $b_1 = (a \vee q) \wedge b$ we have $a \parallel b_1$. Since $b \parallel a_2$ and $b_1 \leqq b$, by (17.11) we have $b_1 <\!| a_2$. Since $p \leqq a_2$ and since $b_1 \parallel a$ and $p \leqq a$, it follows from (18.4.1) that $a = (b_1 \vee p) \wedge a_2 \leqq a_2$. □

Lemma (18.15). *Let l_1 and l_2 be incomplete lines of an affine matroid lattice L. If $l_1 \wedge l_2 \neq 0$, then $l_1 \vee l_2$ is an incomplete element.*

Proof. We may assume $l_1 \neq l_2$. Take a point p with $p \leqq l_1 \wedge l_2$ and two points p_1 and p_2 with $l_i = p \vee p_i$ $(i=1,2)$. Then $p \neq p_1$ and since $l_1 \neq l_2$ we have $p_2 \nleqq l_1 = p \vee p_1$. Hence $\{p, p_1, p_2\}$ is a base of $l_1 \vee l_2$. Since the length of L is not less than 4, there exists a point q with $q \nleqq l_1 \vee l_2$. By (18.11) there exists an element b such that $l_1 \vee l_2 \parallel b$ and $q < b$. Hence $l_1 \vee l_2$ is incomplete by (18.8). □

Lemma (18.16). *Let* $\{a_\alpha; \alpha \in I\}$ *be a finite or infinite set of incomplete elements of an affine matroid lattice L. If* $\bigwedge(a_\alpha; \alpha \in I) \neq 0$ *then* $\bigvee(a_\alpha; \alpha \in I)$ *is an incomplete element.*

Proof. (I) First we shall prove the lemma when $I = \{1,2\}$. Take a point p with $p \leqq a_1 \wedge a_2$. Let l be an arbitrary line contained in $a_1 \vee a_2$. When $p \leqq l$, the three elements l, a_1 and a_2 belong to $L[p,1]$. By (13.2) and the weak modularity of L, $L[p,1]$ is a modular matroid lattice. Since l is a point of $L[p,1]$ and $l \leqq a_1 \vee a_2$, by (9.3.1) there exist points l_1 and l_2 of $L[p,1]$ such that

$$l \leqq l_1 \vee l_2, \quad l_1 \leqq a_1 \quad \text{and} \quad l_2 \leqq a_2.$$

Then l_1 and l_2 are lines of L, and they are incomplete since a_1 and a_2 are incomplete. Since $l_1 \wedge l_2 \geqq p > 0$, by (18.15) $l_1 \vee l_2$ is an incomplete element, whence l is an incomplete line.

When $p \nleqq l$, take points q_1 and q_2 such that $l = q_1 \vee q_2$. Then $p \vee q_1$ is a line which contains p and it is contained in $a_1 \vee a_2$. Hence by the above argument the line $p \vee q_1$ is incomplete. Similarly the line $p \vee q_2$ is incomplete. By (18.15) $(p \vee q_1) \vee (p \vee q_2)$ is an incomplete element which contains $l = q_1 \vee q_2$. Hence l is incomplete. Therefore $a_1 \vee a_2$ is an incomplete element.

(II) If I is a finite set then $\bigvee(a_\alpha; \alpha \in I)$ is an incomplete element by (I).

(III) Let I be an infinite set and let l be an arbitrary line contained in $\bigvee(a_\alpha; \alpha \in I)$. Take points q_1 and q_2 such that $l = q_1 \vee q_2$. Applying (7.12) to both q_1 and q_2, there exists a finite subset J of I such that $l \leqq \bigvee(a_\alpha; \alpha \in J)$. Since $\bigvee(a_\alpha; \alpha \in J)$ is incomplete by (II), l is an incomplete line. Therefore $\bigvee(a_\alpha; \alpha \in I)$ is incomplete. □

Theorem (18.17). *Let L be a non-modular affine matroid lattice. Then for any point p of L there exists a unique greatest incomplete element* $I(p)$ *which contains p. For any two points p and q it holds that either* $I(p) = I(q)$ *or* $I(p) \parallel I(q)$.

Proof. Let $I(p)$ be the join of all incomplete elements containing p. Then by (18.16) $I(p)$ is the greatest incomplete element containing p. When $I(p) \wedge I(q) \neq 0$, by (18.16) $I(p) \vee I(q)$ is an incomplete element which contains both p and q. Hence $I(p) = I(p) \vee I(q) = I(q)$. When $I(p) \wedge I(q) = 0$, since $p \nleqq I(q)$ and $I(q)$ is incomplete, by (18.12) there exists an element a such that $I(q) \parallel a$ and $p \leqq a$. Since by (18.13) a is incomplete, we have $a \leqq I(p)$. It is impossible that $I(q) < I(p)$ since $I(p) \wedge I(q) = 0$. Hence by (17.12) we have $I(q) <\!| I(p)$. Similarly we have $I(p) <\!| I(q)$. Consequently, $I(p) \parallel I(q)$. □

Remark (18.18). Let $\{p, p_\alpha; \alpha \in I\}$ be a base of $I(p)$. Then $I(p) = \bigvee(p \vee p_\alpha; \alpha \in I)$ where $p \vee p_\alpha$ are incomplete lines. Hence we may say that $I(p)$ is the join of all incomplete lines containing p.

Remark (18.19). Let L be an affine matroid lattice. By (18.10), L satisfies the Euclid's strong parallel axiom if and only if 1 is an incomplete element. In this case, $I(p)=1$ for every point $p \in L$.

Remark (18.20). When an affine matroid lattice L is modular, there exists no incomplete element by (18.6). Hence we may put $I(p)=p$ for every point $p \in L$. By (17.2) it holds that either $I(p)=I(q)$ or $I(p) \parallel I(q)$.

EXERCISE 18.1. Let L be an affine matroid lattice. Prove that if L is not modular then there exists a hyperplane having a complement which is not a point.

EXERCISE 18.2. Let a and b be elements of height $\geqq 2$ in an affine matroid lattice. Prove that $a <| b$ if and only if $(b, l)\overline{M}$ for every line l contained in a.

EXERCISE 18.3. Let L be a weakly modular AC-lattice and let $a \neq 1$ in L. Prove that if a contains an incomplete line then there exists a line l such that $l <| a$.

EXERCISE 18.4. Let a be an element of a weakly modular matroid lattice and let $a \neq 1$. Prove that a is modular if and only if every line contained in a is modular.

19. Modular Contractions and Modular Extensions of Affine Matroid Lattices

In this section we shall prove an important result that any affine matroid lattice is a Wilcox lattice.

Lemma (19.1). *In an affine matroid lattice L, if $a <| b$ and $b <| c$ then either $a <| c$ or $a < c$.*

Proof. (I) By (18.8), b is either a point or an incomplete element. When b is a point, by (17.3) a is also a point. Hence, by (17.2), we have either $a \leqq c$ or $a <| c$.

(II) When b is an incomplete element, it follows from (18.14) that there exists a_2 such that $a_2 \parallel b$ and $a \leqq a_2$.

When $a \wedge c \neq 0$, take a point p with $p \leqq a \wedge c$. Since $b <| c$, $p \leqq c$, $b \parallel a_2$ and $p \leqq a_2$, it follows from (18.4.1) that $a_2 = (b \vee p) \wedge c \leqq c$. Hence $a \leqq c$.

When $a \wedge c = 0$, take a point p with $p \leqq a$. We shall show $a \leqq c \vee p$. Since $b <| c \leqq c \vee p$, by (17.12) we have either $b <| c \vee p$ or $b < c \vee p$. In the former case, since $b \| a_2$ and $p \leqq a_2$, it follows from (18.4.1) that $a_2 \leqq c \vee p$, whence $a \leqq c \vee p$. In the latter case, since $a <| b < c \vee p$ and $a \wedge (c \vee p) \geqq p > 0$, it follows from (17.12) that $a < c \vee p$. Since $p \leqq a$ and $p \nleqq c$, in either case we have $a \vee c = c \vee p > c$. Hence $a <| c$. □

Definition (19.2). In a lattice with 0, we write $a \leqq| b$ when either $a <| b$ or $a \leqq b$. We write $a \| | b$ when either $a \| b$ or $a = b$.

Lemma (19.3). *In an affine matroid lattice L,*

(19.3.1) $\quad a \leqq| a$ *for every* $a \in L$,

(19.3.2) $\quad a \leqq| b$ *and* $b \leqq| a$ *imply* $a \|| b$,

(19.3.3) $\quad a \leqq| b$ *and* $b \leqq| c$ *imply* $a \leqq| c$, *and*

(19.3.4) $\quad a \|| b$ *is an equivalence relation.*

Proof. (19.3.1) is evident.

If $a <| b$ and $b \leqq| a$, then since $b \leqq a$ contadicts $a <| b$, we have $b <| a$, and then $a \| b$. If $a \leqq b$ and $b \leqq| a$, then since $b <| a$ contradicts $a \leqq b$, we have $b \leqq a$ and then $a = b$. Thus (19.3.2) holds.

If $a <| b$ and $b <| c$ then $a \leqq| c$ by (19.1). If $a <| b$ and $b \leqq c$ then $a \leqq| c$ by (17.12). If $a \leqq b$ and $b <| c$, then by (17.11) either $a = 0$ or $a <| c$, and then $a \leqq| c$. If $a \leqq b$ and $b \leqq c$ then $a \leqq c$. Thus (19.3.3) holds.

(19.3.4) is a consequence of (19.3.2) and (19.3.3). □

Lemma (19.4). *Let p be a point of an affine matroid lattice L. If a is either an incomplete element or a point of L, then there exists one and only one element b such that*

$$a \|| b \quad \textit{and} \quad p \leqq b.$$

Proof. If a is a point then $b = p$ is the required element. Let a be an incomplete element. When $p \nleqq a$, it follows from (18.12) that there exists one only one element b such that $a \|| b$ and $p \leqq b$. When $p \leqq a$, $b = a$ is the required element. □

Definition (19.5). Let p be a point of an affine matroid lattice L, and let a be either an incomplete element or a point of L. We denote by $\varphi_p(a)$ the element b determined by (19.4), and we call it the $\|$-*image* of a at p, read the parallel image of a at p.

By the proof of (19.4), if $p \leqq a$ then $\varphi_p(a) = a$, and if a is a point then $\varphi_p(a) = p$.

It is easy to show that the domain of the mapping $a \to \varphi_p(a)$ is the set-union $\bigcup(L[q, I(q)]; q \in \Omega(L))$ and its image is $L[p, I(p)]$. If L satisfies the Euclid's strong parallel axiom, then the domain is L with 0 deleted.

Lemma (19.6). *Let p be a point of an affine matroid lattice L, and let each of a and b be either an incomplete element or a point.*

(19.6.1) $\quad a \parallel\!| b$ *if and only if* $\varphi_p(a) = \varphi_p(b)$.

(19.6.2) $\quad a \leqq\!| b$ *if and only if* $\varphi_p(a) \leqq \varphi_p(b)$.

Proof. We need only verify (19.6.2). Since $a \parallel\!| \varphi_p(a)$ and $b \parallel\!| \varphi_p(b)$, it follows from (19.3) that $a \leqq\!| b$ if and only if $\varphi_p(a) \leqq\!| \varphi_p(b)$. But, since $\varphi_p(a) \wedge \varphi_p(b) \geqq p > 0$, $\varphi_p(a) \leqq\!| \varphi_p(b)$ implies $\varphi_p(a) \leqq \varphi_p(b)$. Thus (19.6.2) holds. ▯

Lemma (19.7). *Let p and q be points in an affine matroid lattice L. Two sublattices $L[p, I(p)]$ and $L[q, I(q)]$ are isomorphic by the mutually inverse mappings φ_p and φ_q.*

Proof. It is obvious that $\varphi_p(a) \in L[p, I(p)]$ and $\varphi_q(b) \in L[q, I(q)]$ for $a \in L[q, I(q)]$ and $b \in L[p, I(p)]$. Since $a \parallel\!| \varphi_p(a)$ and $\varphi_p(a) \parallel\!| \varphi_q(\varphi_p(a))$, we have $a \parallel\!| \varphi_q(\varphi_p(a))$. Since $a \wedge \varphi_q(\varphi_p(a)) \geqq q > 0$, we have $a = \varphi_q(\varphi_p(a))$. Similarly we have $\varphi_p(\varphi_q(b)) = b$. Hence the two mappings φ_p and φ_q are mutually inverse one-to-one mappings between $L[p, I(p)]$ and $L[q, I(q)]$. Since φ_p and φ_q preserve the order by (19.6), they are isomorphic. ▯

Remark (19.8). For two points p and q of an affine matroid lattice we have either $I(p) \parallel I(q)$ or $I(p) = I(q)$ by (18.17). When $I(p) \parallel I(q)$, the mappings φ_p and φ_q in (19.7) coincide with the parallel mappings φ and ψ in (17.15). When $I(p) = I(q)$, the mappings φ_p and φ_q are identical on $L[p \vee q, I(p)]$.

Lemma (19.9). *If p is a point of an affine matroid lattice L, then $L[p, I(p)]$ is an irreducible modular matroid sublattice of L.*

Proof. We may assume $p < I(p)$ (see (18.20)). By (13.2), $L[p, I(p)]$ is a matroid sublattice of L, where any point l is a line in L with $p < l \leqq I(p)$. Since L is weakly modular, $L[p, I(p)]$ is modular.

To prove the irreducibility, let $l_1 = p \vee p_1$ and $l_2 = p \vee p_2$ be different points of $L[p, I(p)]$. Then $p \nleqq p_1 \vee p_2$ for otherwise $l_1 = l_2$. Since $p_1 \vee p_2$ is a line of L and since it is contained in $I(p)$, it is an incomplete line. Hence by (18.9) there exists a line l_3 such that $l_3 \parallel p_1 \vee p_2$ and $p < l_3$. By (17.4) we have

$$l_3 \vee p_1 \vee p_2 = p \vee p_1 \vee p_2 = l_1 \vee l_2 .$$

Moreover, $l_3 \neq l_1$ since $p_1 \leqq l_1$ and $p_1 \not\leqq l_3$, and similarly $l_3 \neq l_2$. Hence, in $L[p, I(p)]$, the line $l_1 \vee l_2$ contains a third point l_3. By (16.6.3), $L[p, I(p)]$ is irreducible. □

Definition (19.10). Let L be an affine matroid lattice and let p be a point of L. By the mapping φ_p all incomplete elements and all points of L are transposed into the sublattice $L[p, I(p)]$ preserving the order in the sense of (19.6), and by (19.9) $L[p, I(p)]$ is an irreducible modular matroid lattice. We call $L[p, I(p)]$ a *modular contraction* of L at p.

By (19.7), the modular contractions of L are mutually isomorphic.

Definition (19.11). Let L be an affine matroid lattice. By (19.3.4) $a \parallel\!| b$ is an equivalence relation. When a is an incomplete element of L, we put $[a] = \{b \in L; a \parallel\!| b\}$, and we call $[a]$ an *imaginary element* for L; a is called a *representative* of $[a]$. The set of all imaginary elements of L is denoted by S.

It follows from (18.6) that L is modular if and only if S is empty.

Lemma (19.12). *Let a be an incomplete element of an affine matroid lattice L. If $a \leqq\!| b$ then for any point q with $q \leqq b$ we have $\varphi_q(a) \leqq b$.*

Proof. Since $\varphi_q(a) \parallel\!| a$, by (19.3) we have $\varphi_q(a) \leqq\!| b$. Since $\varphi_q(a) \wedge b \geqq q > 0$, we have $\varphi_q(a) \leqq b$. □

Theorem (19.13). *Let L be a non-modular affine matroid lattice and let S be the set of all imaginary elements for L. In the set $\Lambda \equiv L \cup S$, we define a partial order by the following conventions:*

(19.13.1) *When $a \in L$, $b \in L$ and $a \leqq b$ in L, then $a \leqq b$ in Λ.*

(19.13.2) *When $[a] \in S$, $[b] \in S$ and $a \leqq\!| b$ in L, then $[a] \leqq [b]$ in Λ.*

(19.13.3) *When $[a] \in S$, $b \in L$ and $a \leqq\!| b$ in L, then $[a] < b$ in Λ.*

(19.13.4) *For every element $[a] \in S$, $0 < [a]$ in Λ and there exists no non-zero element $b \in L$ such that $b < [a]$ in Λ.*

Then Λ is an irreducible modular matroid lattice, where the lattice operations $\sqcup$ and $\sqcap$ have the following properties

(19.13.5) *For $a_\alpha \in L\,(\alpha \in I)$, $\bigsqcup(a_\alpha; \alpha \in I) = \bigvee(a_\alpha; \alpha \in I)$.*

(19.13.6) *For $a, b \in L$,*

$$a \sqcap b \begin{cases} = a \wedge b & \text{if } a \wedge b \neq 0 \\ \in S \quad \text{or } = 0 & \text{if } a \wedge b = 0. \end{cases}$$

Moreover

(19.13.7) *$[a] < a$ in Λ for any incomplete element a.*

(19.13.8) *$S = \{\lambda \in \Lambda; 0 < \lambda \leqq [I(p)]\}$ where p is a point of L, and S is isomorphic to the modular contraction $L[p, I(p)]$ with p deleted.*

Proof. (I) It follows from (19.3) that in (19.13.2) (resp. (19.13.3)) the truth of $[a] \leqq [b]$ (resp. $[a] < b$) is independent of the representatives. Since by (18.17) we have $[a] \leqq [I(p)]$ for every $[a] \in S$, it follows from (19.13.4) that $S = \{\lambda \in \Lambda; 0 < \lambda \leqq [I(p)]\}$. By (19.6.2), $[a] \leqq [b]$ if and only if $\varphi_p(a) \leqq \varphi_p(b)$. Hence S is isomorphic to $L[p, I(p)]$ with p deleted, by the mapping $[a] \to \varphi_p(a)$.

(II) We shall show that the join $\bigsqcup(\lambda_\alpha; \alpha \in I)$ in Λ exists for $\lambda_\alpha \in \Lambda (\alpha \in I)$ and that (19.13.5) holds. We may assume $\lambda_\alpha \neq 0$ for every $\alpha \in I$.

(A) When $\lambda_\alpha = a_\alpha \in L$ for every $\alpha \in I$, since by (19.13.4) any upper bound of $\{a_\alpha; \alpha \in I\}$ in Λ is an element of L, the join $\bigvee(a_\alpha; \alpha \in I)$ in L is the join $\bigsqcup(a_\alpha; \alpha \in I)$ in Λ. Hence (19.13.5) holds.

(B) When $\lambda_\alpha = [b_\alpha] \in S$ for every $\alpha \in I$, since S is isomorphic to $L[p, I(p)]$ with p deleted, there exists the join $[c]$ of $\{[b_\alpha]; \alpha \in I\}$ in S (we may take $c = \bigvee(\varphi_p(b_\alpha); \alpha \in I))$). If $d \in L$ is an upper bound of $\{[b_\alpha]; \alpha \in I\}$, then take a point q with $q \leqq d$. Since $b_\alpha \leqq | d$ by (19.13.3), we have $\varphi_q(b_\alpha) \leqq d$ by (19.12). Hence we have $\varphi_q(b_\alpha) \leqq d \wedge I(q)$, whence

$$[b_\alpha] = [\varphi_q(b_\alpha)] \leqq [d \wedge I(q)] < d \quad \text{for every } \alpha \in I.$$

Therefore $[c] < d$ and hence $[c]$ is the join $\bigsqcup([b_\alpha]; \alpha \in I)$ in Λ. We remark that if $\bigwedge(b_\alpha; \alpha \in I) \neq 0$ then

$$\bigsqcup([b_\alpha]; \alpha \in I) = [\bigvee(b_\alpha; \alpha \in I)].$$

(C) For $0 \neq a \in L$ and $[b] \in S$, we shall show that $a \sqcup [b]$ exists. Take a point p with $p \leqq a$. By (19.13.4), any upper bound of a and $[b]$ is an element of L. If $c \in L$ is an upper bound, then since $b \leqq | c$ and $p \leqq a \leqq c$, it follows from (19.12) that $\varphi_p(b) \leqq c$, whence $a \vee \varphi_p(b) \leqq c$. Since $a \vee \varphi_p(b)$ is an upper bound of a and $[b] = [\varphi_p(b)]$, it is the join $a \sqcup [b]$ in Λ. We remark that if $a \wedge b \neq 0$ then $a \sqcup [b] = a \vee b$.

By (A), (B), and (C), the join $\bigsqcup(\lambda_\alpha; \alpha \in I)$ in Λ always exists, and hence Λ is a complete lattice.

(III) We shall prove (19.13.6). When $a \wedge b \neq 0$, take a point p with $p \leqq a \wedge b$, and let λ be a lower bound of a and b in Λ. If $\lambda = c \in L$ then evidently $c \leqq a \wedge b$. If $\lambda = [d] \in S$, then by (19.12) we have $\varphi_p(d) \leqq a$ and $\varphi_p(d) \leqq b$, whence $[d] = [\varphi_p(d)] < a \wedge b$. Hence, $a \wedge b$ is the meet $a \sqcap b$ in Λ.

When $a \wedge b = 0$, by (19.13.4) any lower bound of a and b is either 0 or contained in S. Hence $a \sqcap b \in S$ or $= 0$.

(IV) To prove (19.13.7), let λ be an element of Λ such that $[a] \leqq \lambda \leqq a$. When $\lambda = b \in L$, we have $a \leqq | b \leqq a$. Since $a < | b$ contradicts $b \leqq a$, we have $a \leqq b$, whence $b = a$. When $\lambda = [c] \in S$, since $[c] \leqq a$ implies $[c] \leqq [a]$, we have $[c] = [a]$. Hence $[a] < a$.

(V) We shall show that Λ is atomistic. Any point p of L is a point of Λ, since there is no incomplete element a with $a \leqq | p$. Moreover, for any incomplete line l, the imaginary element $[l]$ is a point of Λ by (19.13.4). Let λ be a non-zero element of Λ. When $\lambda = a \in L$, since L is atomistic, a is a join (in L) of points of L. By (A) of (II), a is a join of points in Λ. When $\lambda = [b] \in S$, let $\{p, p_\alpha; \alpha \in I\}$ be a base of b, and put $l_\alpha = p \vee p_\alpha$ for every $\alpha \in I$. Then $[l_\alpha]$ is a point of Λ. Since $b = \bigvee(l_\alpha; \alpha \in I)$ and $p \leqq l_\alpha$ for every $\alpha \in I$, by (B) of (II) we have $[b] = \bigsqcup([l_\alpha]; \alpha \in I)$. Therefore Λ is atomistic.

(VI) We shall show that Λ is compactly atomistic. It is obvious that a point of Λ is either a point of L or an imaginary element $[l]$ where l is an incomplete line. Let ξ be a point of Λ such that

$$\xi \leqq \bigsqcup(q_\alpha; \alpha \in I) \sqcup \bigsqcup([l_\beta]; \beta \in J),$$

where q_α is a point of L for every $\alpha \in I$ and l_β is an incomplete line for every $\beta \in J$.

When $\xi = p \in L$, I is not empty since $\bigsqcup([l_\beta]; \beta \in J) \in S$ by (B) of (II). Take a point q_0 in $\{q_\alpha; \alpha \in I\}$. We may assume that J is not empty and that $q_0 < l_\beta$ for every $\beta \in J$ (using $\varphi_{q_0}(l_\beta)$ instead of l_β). Putting

$$a = \bigvee(q_\alpha; \alpha \in I) \quad \text{and} \quad b = \bigvee(l_\beta; \beta \in J),$$

by (A) and (B) of (II) we have

$$a = \bigsqcup(q_\alpha; \alpha \in I) \quad \text{and} \quad [b] = \bigsqcup([l_\beta]; \beta \in J).$$

Since $q_0 \leqq a$ and $q_0 \leqq b$, by (C) of (II) we have $a \vee b = a \sqcup [b]$. Hence

$$p \leqq a \vee b = \bigvee(q_\alpha; \alpha \in I) \vee \bigvee(l_\beta; \beta \in J).$$

Since L is compactly atomistic, by (7.12) there exist finite subsets I' and J' of I and J respectively such that $p \leqq \bigvee(q_\alpha; \alpha \in I') \vee \bigvee(l_\beta; \beta \in J')$. We may assume $q_0 \in \{q_\alpha; \alpha \in I'\}$. Putting $a' = \bigvee(q_\alpha; \alpha \in I')$ and $b' = \bigvee(l_\beta; \beta \in J')$, we have

$$p \leqq a' \vee b' = a' \sqcup [b'] = \bigsqcup(q_\alpha; \alpha \in I') \sqcup \bigsqcup([l_\beta]; \beta \in J').$$

When $\xi = [l] \in S$ and I is not empty, we take a point q_0 in $\{q_\alpha; \alpha \in I\}$, and we may assume $q_0 < l$. As above we have $[l] \leqq a \vee b$, whence $l \leqq | a \vee b$. Since $q_0 \leqq l \wedge (a \vee b)$, we have $l \leqq a \vee b$. Applying (7.13) to two points p_1 and p_2 with $p_1 \vee p_2 = l$, there exist finite subsets I' and J' such that

$$l \leqq \bigvee(q_\alpha; \alpha \in I') \vee \bigvee(l_\beta; \beta \in J').$$

Then $[l] < l \leqq \bigsqcup(q_\alpha; \alpha \in I') \sqcup \bigsqcup([l_\beta]; \beta \in J')$.

When $\xi = [l] \in S$ and I is empty, we take a point q_0 of L with $q_0 < l$. Then, as above, we have $[l] \leqq [b]$, whence $l \leqq b$. Hence there exists a finite subset J' such that $l \leqq \bigvee(l_\beta; \beta \in J')$, whence

$$[l] \leqq \bigsqcup([l_\beta]; \beta \in J').$$

(VII) We shall show that Λ has the covering property. Let $0 \neq \lambda \in \Lambda$ and let ξ be a point of Λ with $\xi \not\leqq \lambda$.

When $\lambda = a \in L$ and $\xi = p \in L$, then $a \sqcup p = a \vee p$. If $a \leqq \gamma \leqq a \vee p$ in Λ, then $\gamma \in L$ since $0 \neq a \in L$. Hence, by the covering property of L, $\gamma = a$ or $\gamma = a \vee p$. Therefore $a \lessdot a \sqcup p$ in Λ.

When $\lambda = a \in L$ and $\xi = [l] \in S$, taking a point $p \in L$ with $p \leqq a$ we have $a \sqcup [l] = a \vee \varphi_p(l)$ by (C) of (II). We take a point $q \in L$ such that $\varphi_p(l) = p \vee q$. Since $[l] \not\leqq a$, we have $q \not\leqq a$. Hence $a \sqcup [l] = a \vee p \vee q = a \vee q \gtrdot a$ in L. As above we have $a \lessdot a \sqcup [l]$ in Λ.

When $\lambda = [b] \in S$ and $\xi = p \in L$, then $[b] \sqcup p = \varphi_p(b) \vee p = \varphi_p(b)$. By (19.13.7) we have

$$[b] = [\varphi_p(b)] \lessdot \varphi_p(b) = [b] \sqcup p.$$

When $\lambda \in S$ and $\xi \in S$, λ and ξ belong to the sublattice $\Lambda[0, [I(p)]]$. By (19.13.8) this sublattice is isomorphic to the modular contraction $L[p, I(p)]$ which has the covering property. Hence we have $\lambda \lessdot \lambda \sqcup \xi$.

Thus Λ has the covering property and hence it is a matroid lattice.

(VIII) We shall show that Λ has the dual covering property. Let $1 \neq \lambda \in \Lambda$ and let π be a hyperplane of Λ with $\lambda \not\leqq \pi$. We remark that if $\pi = h \in L$ then h is a hyperplane of L, and that if $\pi = [b] \in S$ then $b = 1$ since $[b] < b$ and $[b] \lessdot 1$.

When $\lambda = a \in L$, $\pi = h \in L$ and $a \wedge h \neq 0$, by (19.13.6) we have $a \sqcap h = a \wedge h$. If $a \wedge h \leqq \gamma \leqq a$ in Λ then $\gamma = c \in L$. Since L is weakly modular, $L[a \wedge h, 1]$ is modular. Hence $a \not\leqq h$ implies $a \wedge h \lessdot a$ in L. Therefore $c = a \wedge h$ or $c = a$, whence $a \sqcap h \lessdot a$ in Λ.

When $\lambda = a \in L$, $\pi = h \in L$ and $a \wedge h = 0$, we have $a <| h$ in L since $h \lessdot 1 = a \vee h$. Hence a is either a point or an incomplete element in L by (18.8). If a is a point of L, then since a is a point of Λ and since $a \not\leqq h$, we have

$$a \sqcap h = 0 \lessdot a.$$

If a is an incomplete element, then since $[a] < h$ and $[a] < a$, we have $[a] \leqq a \sqcap h < a$. Since $[a] \lessdot a$ by (19.13.7), it holds that $a \sqcap h \lessdot a$.

When $\lambda = a \in L$ and $\pi = [1] \in S$, since 1 is an incomplete element, a is either a point or an incomplete element. If a is a point then $a \sqcap [1] = 0 \lessdot a$. If a is an incomplete element, then since $[a] \leqq a \sqcap [1] < a$ and $[a] \lessdot a$, we have $a \sqcap [1] \lessdot a$.

When $\lambda=[b]\in S$, then $\pi=h\in L$, since otherwise $\pi=[1]\geqq[b]=\lambda$, a contradiction. Let $[b]\sqcap h<\gamma\leqq[b]$ in Λ. Then $\gamma=[c]\in S$. Take a point p of L with $p\leqq h$. We may assume $p\leqq b$ and $p\leqq c$. Since $c\leqq| b$ and $b\wedge c\geqq p$, we have $c\leqq b$. Now we shall show $b\wedge h<c$. If $b\wedge h=p$, this is obvious. If $b\wedge h>p$, then $b\wedge h$ is an incomplete element and then

$$[b\wedge h]\leqq[b]\sqcap h<[c].$$

Since $b\wedge h\wedge c\geqq p$, it holds that $b\wedge h<c$. Therefore we have

$$p\leqq b\wedge h<c\leqq b.$$

Since $L[p,1]$ is modular, we have $b\wedge h\lessdot b$ in L. Hence $c=b$, whence $\gamma=[b]$. Therefore $[b]\sqcap h\lessdot[b]$.

Thus Λ has the dual covering property and hence it is a modular matroid lattice by (14.1).

(IX) Finally we shall show that Λ is irreducible. It suffices to show that the join of two different points ξ and η contains a third point in Λ.

When $\xi=p\in L$ and $\eta=q\in L$, the join $p\sqcup q$ is a line $p\vee q$ in L. If it is a complete line, then it contains a third point by (18.7). If it is an incomplete line, then the imaginary element $[p\vee q]$ is a point of Λ and $[p\vee q]<p\vee q$. Hence $p\sqcup q$ always contains a third point.

When $\xi=p\in L$ and $\eta=[l]\in S$, we have $p\sqcup[l]=\varphi_p(l)$ by (C) of (II). Since $\varphi_p(l)$ is a line of L, there exists a point q of L such that $\varphi_p(l)=p\vee q$. Evidently q is a third point contained in $p\sqcup[l]$.

When $\xi=[l]\in S$ and $\eta=[k]\in S$, we take a point p of L with $p\leqq l$ and we may assume $p\leqq k$. Moreover we take two points q and r of L such that $l=p\vee q$ and $k=p\vee r$. We have $q\neq r$, $q\vee r\neq l$ and $q\vee r\neq k$, since otherwise $l=k$, whence $\xi=\eta$, a contradiction. Since $q\vee r\leqq l\vee k \leqq I(p)$, $q\vee r$ is an incomplete line and we have

$$[q\vee r]\leqq[l\vee k]=[l]\sqcup[k].$$

If $[q\vee r]=[l]$, then since $(q\vee r)\wedge l\geqq q$, we would have $q\vee r=l$, a contradiction. Hence $[q\vee r]\neq[l]$. Similarly $[q\vee r]\neq[k]$. Thus $[q\vee r]$ is a third point contained in $[l]\sqcup[k]$. □

Corollary (19.14). *If L is a non-modular affine matroid lattice then L is a Wilcox lattice with imaginary unit, where its modular extension is the lattice Λ given by* (19.13). (The set of imaginary elements in the sense of (3.12) coincides with the set of imaginary elements in the sense of (19.11).)

Proof. The lattice Λ in (19.13) is a complemented modular lattice. By (19.13.8), the subset S of Λ has the properties (3.11.1) and (3.11.2). Moreover, by (19.13.1), L has the same order as Λ. Hence L is a Wilcox lattice. Evidently L has the imaginary unit $[I(p)]$, and S is the set of imaginary elements in the sense of (3.12). □

EXERCISE 19.1. Let L be a non-modular affine matroid lattice and let $a \wedge b = 0$ in L. Prove that if $a \wedge (p \vee b) > p$ for some point $p \leqq a$ then $a \sqcap b = [a \wedge (p \vee b)]$ in Λ.

20. Atomistic Wilcox Lattices

Remark (20.1). A Wilcox lattice $L \equiv \Lambda - S$ is M-symmetric by (3.11). Hence, if L is atomistic then it is an AC-lattice by (7.7), and if L is compactly atomistic then it is a matroid lattice.

Lemma (20.2). *Let $L \equiv \Lambda - S$ be a Wilcox lattice.*

(20.2.1) *For two elements a and b of L, $a \lessdot b$ in L if and only if $a \lessdot b$ in Λ.*

(20.2.2) *For an element p of L, p is a point of L if and only if p is a point of Λ.*

(20.2.3) *If p is a point of L then $p \sqcap u = 0$ for every $u \in S$.*

Proof. If $a \lessdot b$ in Λ then $a \lessdot b$ in L since L is a subset of Λ. Conversely assume $a \lessdot b$ in L. Let λ be an element of Λ such that $a \leqq \lambda \leqq b$. When $\lambda \in L$, we have $\lambda = a$ or $\lambda = b$, by the assumption. When $\lambda \in S$, since Λ is relatively complemented, there exists $\lambda' \in \Lambda$ such that

$$\lambda \sqcup \lambda' = b \quad \text{and} \quad \lambda \sqcap \lambda' = a.$$

If $\lambda' \in S$ then by (3.11.2) we would have $b = \lambda \sqcup \lambda' \in S$, a contradiction. Hence $\lambda' \in L$. Since $a \leqq \lambda' \leqq b$, we have $\lambda' = a$ or $\lambda' = b$, whence $\lambda = b$ or $\lambda = a$. Therefore $a \lessdot b$ in Λ. This completes the proof of (20.2.1). (20.2.2) and (20.2.3) follow from (20.2.1). □

Lemma (20.3). *Let $L \equiv \Lambda - S$ be a Wilcox lattice. If L is atomistic then so is Λ.*

Proof. (I) Let $0 \neq a \in L$. Since any upper bound λ of the set $\{p \in \Omega(L); p \leqq a\}$ in Λ belongs to L, we have

$$\lambda \geqq \bigvee (p \in \Omega(L); p \leqq a) = a.$$

Hence a is the join $\bigsqcup (p \in \Omega(L); p \leqq a)$ in Λ and here p is a point of Λ by (20.2.2).

(II) Let $u \in S$. We take an arbitrary point p of L. Since $p \sqcap u = 0$ by (20.2.3), by the modularity of Λ we have $\Lambda[0,u] \cong \Lambda[p, p \sqcup u]$. Since $\Lambda[p, p \sqcup u]$ is included in L, we have

$$\Lambda[0,u] \cong L[p, p \sqcup u].$$

Since L is an AC-lattice, the sublattice $L[p, p \sqcup u]$ is atomistic by (8.18). Hence $\Lambda[0, u]$ is atomistic, whence u is the join of points of Λ. □

Remark (20.4). Let $L \equiv \Lambda - S$ be an atomistic Wilcox lattice, and let a be an element of L. It follows from (20.2) and the proof (I) of (20.3) that a is finite in L if and only if a is finite in Λ. Moreover, the height of a in L coincides with that in Λ.

Remark (20.5). Let $L \equiv \Lambda - S$ be an atomistic Wilcox lattice of finite length. Then L is a matroid lattice by (8.10). Moreover, by (20.3) and (20.4), Λ is atomistic and of finite length, whence Λ is a modular matroid lattice. If S is not empty then there exists the imaginary unit for L, because S has an element with a maximal height, which must be a greatest element in S.

Lemma (20.6). *Let $L \equiv \Lambda - S$ be an atomistic Wilcox lattice of length* $\geqq 3$ (may be of infinite length). *L is modular if and only if S is empty.*

Proof. If S is empty then evidently L is modular. If S is not empty, then there exists a point s in S and exists a point p in L, and then $l = p \sqcup s$ is a line in L. Since the length of $L \geqq 3$, there exists a point q of L such that $q \nleqq l$. Since Λ is modular and since $q \sqcap l = 0$, we have

$$(s \sqcup q) \sqcap l = s \sqcup (q \sqcap l) = s \in S,$$

whence $(s \sqcup q, l)\overline{M}$ by (3.11.5). Hence L is not modular. □

Lemma (20.7). *In a Wilcox lattice $L \equiv \Lambda - S$, let a be an element which is neither zero nor a point. Then $a <\!| b$ in L if and only if*

(20.7.1) $\quad a \sqcap b \in S$ *and* $a \sqcap b \lessdot a$ *in* Λ.

Proof. Since Λ is a modular lattice,

(1) $\quad b \lessdot a \sqcup b$ in Λ if and only if $a \sqcap b \lessdot a$ in Λ.

If $a <\!| b$ in L, then by (17.6.1) we have $(b, a)\overline{M}$, whence $a \sqcap b \in S$ by (3.11.5). Moreover by (20.2.1) we have $b \lessdot a \vee b = a \sqcup b$ in Λ. Hence $a \sqcap b \lessdot a$ in Λ by (1). Conversely, if (20.7.1) holds, then by (1) $b \lessdot a \sqcup b$ in Λ, whence $b \lessdot a \vee b$ in L. Moreover $a \sqcap b \in S$ implies $a \wedge b = 0$ by (3.11.4). Hence $a <\!| b$ in L. □

Lemma (20.8). *Let $L \equiv \Lambda - S$ be an atomistic Wilcox lattice and let l be a line of L. If, in Λ, l contains a point in S then for any point p of L with $p \nleqq l$ there exists one and only one line k such that $l \parallel k$ and $p < k$. If l contains no point in S then there exists no line parallel to l.*

Proof. By (20.3) and (20.4), Λ is an atomistic modular lattice and $h(l) = 2$ in Λ.

Assume that l contains a point $s \in S$. For a point p of L with $p \nleq l$, we put $k = p \sqcup s$. In Λ, since $h(l) = h(k) = 2$ and $l \neq k$, we have $h(l \sqcap k) \leqq 1$. Since $s \leqq l \sqcap k$, it holds that $h(l \sqcap k) = 1$ and $l \sqcap k = s$, whence

$$l \sqcap k \in S, \quad l \sqcap k \lessdot l \text{ and } l \sqcap k \lessdot k \quad \text{in } \Lambda.$$

By (20.7) we have $l \parallel k$ in L.

We shall show the uniqueness of k. If $l \parallel d$ and $p < d$, then since $l > l \sqcap d \in S$ by (20.7), $t = l \sqcap d$ is a point in S. We have $d = p \sqcup t$ since $t \lessdot d$. If $s \neq t$, then since $s \leqq l$ and $t \leqq l$, we would have $l = s \sqcup t \in S$, a contradiction. Hence $s = t$, whence $d = k$.

Assume that l contains no point in S. If we had a line k with $l \parallel k$, then as above $l \sqcap k$ would be a point in S, a contradiction. Hence there is no element parallel to l. □

Theorem (20.9). *If a Wilcox lattice $L \equiv \Lambda - S$ is a matroid lattice of length $\geqq 4$ then L is an affine matroid lattice.*

Proof. L is weakly modular by (3.11), and L satisfies Euclid's weak parallel axiom by (20.8). Hence L is an affine matroid lattice. □

Corollary (20.10). *A weakly modular matroid lattice L of length $\geqq 4$ is a Wilcox lattice if and only if L satisfies Euclid's weak parallel axiom.*

Proof. This follows from (20.9) and (19.14). □

Lemma (20.11). *Let $L \equiv \Lambda - S$ be an atomistic Wilcox lattice. L is complete if and only if Λ is complete.*

Proof. (I) If Λ is complete then so is L by (3.15).

(II) Assume that L is complete. Let $a_\alpha \in L$ for every $\alpha \in I$. We shall show that the join $\bigsqcup(a_\alpha; \alpha \in I)$ in Λ exists. If $a_\alpha = 0$ for every α, then $\bigsqcup(a_\alpha; \alpha \in I) = 0$. If $a_\alpha \neq 0$ for some $\alpha \in I$, then any upper bound of $\{a_\alpha; \alpha \in I\}$ in Λ belongs to L. Hence the join $\bigvee(a_\alpha; \alpha \in I)$ in L is the join $\bigsqcup(a_\alpha; \alpha \in I)$ in Λ.

(III) Let $u_\alpha \in S$ for every $\alpha \in I$. When $\{u_\alpha\}$ has no upper bound except 1, we have $\bigsqcup_\alpha u_\alpha = 1$. When $\{u_\alpha\}$ has an upper bound λ with $\lambda < 1$, we can take a non-zero element a in L such that $a \sqcap \lambda = 0$; because if $\lambda \in L$, then there exists a point p of L with $p \nleq \lambda$, whence $p \sqcap \lambda = 0$, and if $\lambda \in S$, then taking a complement a of λ in Λ, we have $a \neq 0$ and $a \in L$.

Since $u_\alpha \sqcup a \in L$ for every $\alpha \in I$, there is the join $b = \bigvee(u_\alpha \sqcup a; \alpha \in I)$ in L. By the modularity of Λ

$$(u_\alpha \sqcup a) \sqcap \lambda = u_\alpha \sqcup (a \sqcap \lambda) = u_\alpha,$$

whence $u_\alpha \leqq b \sqcap \lambda$ for every α. If $\bar{\lambda}$ is an arbitrary upper bound of $\{u_\alpha\}$ in Λ, then putting $c = (\bar{\lambda} \sqcap \lambda) \sqcup a$, we have $c \in L$ and $u_\alpha \sqcup a \leqq c$ for every α, whence $b \leqq c$. Hence

$$b \sqcap \lambda \leqq c \sqcap \lambda = \{(\bar{\lambda} \sqcap \lambda) \sqcup a\} \sqcap \lambda = (\bar{\lambda} \sqcap \lambda) \sqcup (a \sqcap \lambda) = \bar{\lambda} \sqcap \lambda \leqq \bar{\lambda}.$$

Therefore $b \sqcap \lambda$ is the join $\bigsqcup(u_\alpha; \alpha \in I)$.

(IV) By (II) and (III), it is easy to show that any subset of Λ has its join in Λ. Hence Λ is complete. □

Theorem (20.12). (Parallelism and modularity). *Let* $L \equiv \Lambda - S$ *be a complete atomistic Wilcox lattice with imaginary unit* i (in particular, let L be a non-modular affine matroid lattice), *and let* a *and* b *be non-zero elements of* L. *The following three statements are equivalent.*

(α) $a <\!| \, b$.

(β) $(b, a_1)\bar{M}$ *for every element* a_1 *such that* $a_1 \leqq a$ *and* $h(a_1) \geqq 2$.

(γ) $(b, l)\bar{M}$ *for every line* l *such that* $l \leqq a$.

Proof. The implication $(\alpha) \Rightarrow (\beta)$ follows from (17.6.1), and $(\beta) \Rightarrow (\gamma)$ is trivial.

$(\gamma) \Rightarrow (\alpha)$. ($\alpha$) holds if $h(a)=1$. If $h(a) \geqq 2$, then by (8.16) a has a base $\{p, p_\alpha; \alpha \in I\}$ where I is not empty. For every $\alpha \in I$, since $(b, p \vee p_\alpha)\bar{M}$ by (γ), putting $u_\alpha = b \sqcap (p \vee p_\alpha)$, we have $u_\alpha \in S$ by (3.11.5). Since $p < p \sqcup u_\alpha \leqq p \vee p_\alpha = p \sqcup p_\alpha$, we have $p \sqcup u_\alpha = p \sqcup p_\alpha$. Since Λ is complete by (20.11),

$$a = p \vee \bigvee_\alpha p_\alpha = p \sqcup \bigsqcup_\alpha p_\alpha = p \sqcup \bigsqcup_\alpha u_\alpha .$$

We have $\bigsqcup_\alpha u_\alpha \in S$, since $\{u_\alpha\}$ has an upper bound $i \in S$. If we had $a \leqq b$, then taking a line $l \leqq a$, we would have $(b, l)M$, a contradiction. Hence $\bigsqcup_\alpha u_\alpha \leqq a \sqcap b < a$. Since $\bigsqcup_\alpha u_\alpha < p \sqcup \bigsqcup_\alpha u_\alpha = a$ in Λ, we have

$$a \sqcap b = \bigsqcup_\alpha u_\alpha \in S \quad \text{and} \quad a \sqcap b \lessdot a \quad \text{in } \Lambda .$$

Therefore $a <\!| \, b$ by (20.7). □

(The equivalence of (α) and (β) can be proved whenever L is an atomistic Wilcox lattice. See (24.2).)

Lemma (20.13). *Let* $L \equiv \Lambda - S$ *be a Wilcox lattice with imaginary unit* i.

(20.13.1) *If* Λ *is atomistic then* L *is atomic.*

(20.13.2) *If* Λ *is compactly atomistic and irreducible then* L *is compactly atomistic.*

Proof. (I) Assume that Λ is atomistic. We remark that an element λ of Λ belongs to S if $0 < \lambda \leqq i$. Let a be a non-zero element of L. Since $a \not\leqq i$ and since Λ is atomistic, there exists a point p of Λ such that $p \leqq a$ and $p \not\leqq i$. Then p is a point of L with $p \leqq a$.

(II) Assume that Λ is compactly atomistic and irreducible. L is complete by (3.15) and is atomic by (I). We shall show that L is atomistic. Let a be a non-zero element of L and let ω be the set of points p of L with $p \leqq a$. Since L is atomic, ω is not empty. Put $a_1 = \bigvee(p; p \in \omega)$. Since Λ is atomistic, in order to prove $a = a_1$ it suffices to show that if

s is a point of Λ with $s \leqq a$ then $s \leqq a_1$. When $s \in L$, then $s \in \omega$, whence $s \leqq a_1$. When $s \in S$, take a point $p \in \omega$. It follows from (16.6.3) that the line $p \sqcup s$ contains a third point q in Λ, and then

$$p \sqcup q = p \sqcup s = s \sqcup q.$$

We have $q \in L$, since otherwise $p \leqq s \sqcup q \in S$, a contradiction. Since $q \leqq p \sqcup s \leqq a$, we have $q \in \omega$. Hence

$$s \leqq p \sqcup q \leqq a_1.$$

Therefore $a = a_1 = \bigvee(p; p \in \omega)$.

Next, we shall show that L is compactly atomistic. If p and $q_\alpha (\alpha \in I)$ are points of L and $p \leqq \bigvee(q_\alpha; \alpha \in I)$, then by (20.2.2) and (3.15.1) p and q_α are points of Λ and $p \leqq \bigsqcup(q_\alpha; \alpha \in I)$. Since Λ is compactly atomistic, there exists a finite subset J of I such that

$$p \leqq \bigsqcup(q_\alpha; \alpha \in J) = \bigvee(q_\alpha; \alpha \in J). \quad \square$$

Remark (20.14). If Λ is atomistic, then L is not necessarily atomistic. The third lattice in Exercise 1.1 is not atomistic though it has an atomistic modular extension.

Theorem (20.15). *Let Λ be an irreducible modular matroid lattice of length $\geqq 4$. For any element $i \in \Lambda$ with $0 < i < 1$, the Wilcox lattice $L \equiv \Lambda - S$ with imaginary unit i is a non-modular affine matroid lattice. L satisfies Euclid's strong parallel axiom if and only if i is a hyperplane of Λ.*

Proof. (I) L is compactly atomistic by (20.13.2) and hence it is a matroid lattice. By (20.4) L is of length $\geqq 4$. Hence it follows from (20.9) that L is an affine matroid lattice. L is not modular by (20.6).

(II) It follows from (20.8) that L satisfies Euclid's strong parallel axiom if and only if every line of L contains a point in S.

If i is a hyperplane of Λ, then for any line l of L, since $i \sqcup l = 1 > i$, we have $i \sqcap l < l$. Hence $i \sqcap l$ is a point in S.

If i is not a hyperplane of Λ, then a complement of i contains a line l of Λ. Since $l \sqcap i = 0$, we have $l \in L$, and l contains no point in S.

Therefore L satisfies Euclid's strong parallel axiom if and only if i is a hyperplane of Λ. $\square$

Exercise 20.1. Let $L \equiv \Lambda - S$ be a Wilcox lattice. Prove that if Λ is upper continuous and if L is atomic then S has the imaginary unit.

Exercise 20.2. Let H be a Hilbert space (infinite dimensional). Let Λ be the set of closed subspaces M such that either M or its orthocomplement $M^\perp$ is finite dimensional, and let S be the set of non-zero, finite dimensional subspaces. Prove that Λ is a complemented modular lattice which is not complete, ordered by set-inclusion, and that $L \equiv \Lambda - S$ is a Wilcox lattice which is complete.

EXERCISE 20.3. Let $L \equiv \Lambda - S$ be a Wilcox lattice with imaginary unit. Prove that if L is complete then so is Λ.

EXERCISE 20.4. Prove that any atomistic Wilcox lattice is left complemented.

21. Singular Elements in Atomistic Wilcox Lattices

Definition (21.1). An element a of a Wilcox lattice $L \equiv \Lambda - S$ is called a *singular element* when there exist a point p of L and an element u of S such that $a = p \sqcup u$. u is called an *imaginary part* of a and we write $u = \imath(a)$.

Lemma (21.2). *Let a be a singular element of a Wilcox lattice $L \equiv \Lambda - S$. An imaginary part $\imath(a)$ of a is uniquely determined by a, and it is the greatest element of S contained in a. Moreover, $\imath(a) \lessdot a$ in Λ.*

Proof. Let $a = p \sqcup u$, $p \in \Omega(L)$ and $u \in S$. If $v \in S$ and $v \leqq a$, then since $u \sqcup v \in S$, we have $p \sqcap (u \sqcup v) = 0$ by (20.2.3). Hence by the modularity of Λ

$$u = u \sqcup \{p \sqcap (u \sqcup v)\} = (u \sqcup p) \sqcap (u \sqcup v) = a \sqcap (u \sqcup v) \geqq v.$$

Therefore u is the greatest element of S contained in a. By the covering property of Λ we have $u \lessdot p \sqcup u = a$. ▯

Lemma (21.3). *Let a be a singular element of a Wilcox lattice $L \equiv \Lambda - S$, and a_1 be a non-zero element of L with $a_1 \leqq a$.*

(21.3.1) $a = a_1 \sqcup \imath(a)$.

(21.3.2) *If a_1 contains a point of L then a_1 is either a point or a singular element with the imaginary part $\imath(a_1) = a_1 \sqcap \imath(a)$.*

Proof. Since $\imath(a) < a_1 \sqcup \imath(a) \leqq a$ and since $\imath(a) \lessdot a$ by (21.2), (21.3.1) holds.

If a_1 contains a point p of L then by (21.3.1) we have $a = p \sqcup \imath(a)$. Hence

$$a_1 = (p \sqcup \imath(a)) \sqcap a_1 = p \sqcup (\imath(a) \sqcap a_1).$$

If $p < a_1$, then $\imath(a) \sqcap a_1 \neq 0$, whence $\imath(a) \sqcap a_1 \in S$. Hence a_1 is a singular element with the imaginary part $\imath(a) \sqcap a_1$. ▯

Lemma (21.4). *In a Wilcox lattice $L \equiv \Lambda - S$, let a be an element which is neither zero nor a point. Then, $a <| b$ in L if and only if a is a singular element and $\imath(a) = a \sqcap b$.*

Proof. It follows from (20.7) that $a<|b$ in L if and only if

(1) $$a \sqcap b \in S \quad \text{and } a \sqcap b \lessdot a \quad \text{in } \Lambda.$$

If a is singular and $\iota(a) = a \sqcap b$ then (1) holds by (21.2). Conversely, if (1) holds then we take a complement p of $a \sqcap b$ in $\Lambda[0,a]$. Since $\Lambda[0,p] \cong \Lambda[a \sqcap b, a]$ by the modularity of Λ, p is a point of Λ. Moreover we have $p \in L$, since otherwise $a = p \sqcup (a \sqcap b) \in S$, a contradiction. Hence a is a singular element and $\iota(a) = a \sqcap b$. □

Lemma (21.5). *Let a and b be two singular elements of a Wilcox lattice $L \equiv \Lambda - S$.*

(21.5.1) $$\iota(a) \leqq \iota(b) \quad \textit{if and only if} \quad a \leqq| b.$$

(21.5.2) $$\iota(a) = \iota(b) \quad \textit{if and only if} \quad a \||| b.$$

Proof. Let $a = p \sqcup \iota(a)$ and $b = q \sqcup \iota(b)$. Assume $\iota(a) \leqq \iota(b)$. If $p \leqq b$ then we have

$$a = p \sqcup \iota(a) \leqq b \sqcup \iota(b) = b.$$

If $p \nleqq b$, then since $\iota(a) \leqq b$ and $p \sqcap b = 0$, we have

$$a \sqcap b = (\iota(a) \sqcup p) \sqcap b = \iota(a) \sqcup (p \sqcap b) = \iota(a).$$

Hence $a<|b$ by (21.4).

Conversely, assume $a \leqq| b$. When $a \leqq b$, we have $\iota(a) \leqq \iota(b)$ by (21.2). When $a<|b$, by (21.4) we have $\iota(a) = a \sqcap b \leqq b$. Hence $\iota(a) \leqq \iota(b)$ by (21.2). This completes the proof of (21.5.1). (21.5.2) follows from (21.5.1). □

Theorem (21.6). *In a Wilcox lattice $L \equiv \Lambda - S$, if a is a singular element, then for any point $q \in L$ with $q \nleqq a$ there exists one and only one element b such that $a \| b$ and $q < b$.*

Proof. Let $a = p \sqcup \iota(a)$ where $p \in \Omega(L)$. For any $q \in \Omega(L)$ with $q \nleqq a$, we put $b = q \sqcup \iota(a)$. Then b is a singular element and $\iota(b) = \iota(a)$. Hence we have $a \||| b$ by (21.5.2). But $a \neq b$ since $q \nleqq a$ and $q \leqq b$. Therefore we have $a \| b$.

If c is an element such that $a \| c$ and $q < c$, then c is singular by (21.4), and $\iota(c) = \iota(a)$ by (21.5.2). Hence by (21.3.1) we have

$$c = q \sqcup \iota(c) = q \sqcup \iota(a) = b.$$

Therefore b is unique. □

Lemma (21.7). *Let $L \equiv \Lambda - S$ be an atomistic Wilcox lattice of length $\geqq 3$.*

(21.7.1) *If a is an element of L with $h(a) \geqq 2$ then the following three statements are equivalent.*

(α) *a is singular and $a \neq 1$.*
(β) *There exists an element $b \in L$ with $a \| b$.*
(γ) *There exists an element $b \in L$ with $a <| b$.*

(21.7.2) *$1 \in L$ is a singular element if and only if there exist two singular elements a_1 and a_2 such that*

$$a_i \neq 1 \quad (i=1,2), \qquad a_1 \vee a_2 = 1 \quad \textit{and} \quad a_1 \wedge a_2 \neq 0.$$

Proof. (I) First we shall prove (21.7.1). Since L is atomistic, if $a \neq 1$ then there exists a point $p \in L$ such that $p \nleqq a$. Hence, it follows from (21.6) that (α) implies (β). The implication (β)$\Rightarrow$(γ) is trivial, and the implication (γ)$\Rightarrow$(α) follows from (21.4).

(II) Assume that 1 is a singular element. Then $\imath(1)$ is a hyperplane, since $\imath(1) \lessdot 1$ by (21.2). Since L is of length $\geqq 3$, so is Λ. Hence there is an element u_1 with $0 < u_1 < \imath(1)$. Let u_2 be a complement of u_1 in $\Lambda[0, \imath(1)]$, and we put $a_i = p \sqcup u_i$ $(i=1,2)$. Since $u_i \in S$ and $u_i < \imath(1)$, a_1 and a_2 are singular and $a_i \neq 1$. Moreover,

$$a_1 \vee a_2 = a_1 \sqcup a_2 = p \sqcup u_1 \sqcup u_2 = p \sqcup \imath(1) = 1 \quad \text{and} \quad a_1 \wedge a_2 \geqq p > 0.$$

(III) Assume that a_1 and a_2 are singular elements with $a_i \neq 1$, $a_1 \vee a_2 = 1$ and $a_1 \wedge a_2 \neq 0$. Take a point $p \in L$ with $p \leqq a_1 \wedge a_2$. Then $a_i = p \sqcup \imath(a_i)$ by (21.3.1). Hence

$$1 = a_1 \vee a_2 = a_1 \sqcup a_2 = p \sqcup \imath(a_1) \sqcup \imath(a_2).$$

Since $\imath(a_1) \sqcup \imath(a_2) \in S$, 1 is singular. □

Remark (21.8). In an atomistic Wilcox lattice L of length $\geqq 3$, it follows from (21.7) that we can define singular elements without using imaginary elements. We may say that an element $a \in L$ with $a \neq 1$ and $h(a) \geqq 2$ is singular when there exists an element parallel to a, and we may say that 1 is singular when there exist two singular elements a_1 and a_2 such that $a_i \neq 1$ $(i=1,2)$, $a_1 \vee a_2 = 1$ and $a_1 \wedge a_2 \neq 0$.

Theorem (21.9). *Let L be an atomistic Wilcox lattice of length $\geqq 3$. The modular extension Λ of L is uniquely determined up to isomorphism.*

Proof. (I) Let $L \equiv \Lambda_1 - S_1$ and $L \equiv \Lambda_2 - S_2$. If L is modular, then by (20.6) we have $\Lambda_1 \equiv \Lambda_2 \equiv L$. Hence we may assume that L is not modular. Then S_1 and S_2 are not empty. We shall show that there exists an isomorphism of Λ_1 onto Λ_2 which is the identity on L. It follows from (21.8) that for any singular element a of L there are two imaginary parts $\iota_1(a)\in S_1$ and $\iota_2(a)\in S_2$. We take a point p of L, and for $u\in S_1$ and $v\in S_2$ we define

$$\varphi(u)=\iota_2(p\sqcup_1 u) \quad \text{and} \quad \psi(v)=\iota_1(p\sqcup_2 v),$$

where $\sqcup_i$ denotes the join in Λ_i $(i=1,2)$.

(II) For two points p and q of L, we put $a=p\sqcup_1 u$ and $b=q\sqcup_1 u$. Since $\iota_1(a)=\iota_1(b)=u$, we have $a\parallel\!\mid b$ by (21.5.2). Hence $\iota_2(a)=\iota_2(b)$ by (21.5.2). Therefore the mapping $\varphi: S_1\to S_2$ is determined independent from the choice of p, as is the mapping $\psi: S_2\to S_1$. Hence

(1) $p\sqcup_2\varphi(u)=p\sqcup_1 u$ and $p\sqcup_1\psi(v)=p\sqcup_2 v$ for every point p of L.

(III) For $u\in S_1$ it follows from (1) that

$$\psi\,\varphi(u)=\iota_1(p\sqcup_2\varphi(u))=\iota_1(p\sqcup_1 u)=u.$$

Similarly we have $\varphi\psi(v)=v$ for $v\in S_2$. Hence φ and ψ are mutually inverse one-to-one mappings between S_1 and S_2, and evidently there are order-preserving.

(IV) We define a mapping $\overline{\varphi}$ of Λ_1 into Λ_2 as follows:

$$\overline{\varphi}(\lambda) = \begin{cases} \varphi(\lambda) & \text{when } \lambda\in S_1, \\ \lambda & \text{when } \lambda\in L. \end{cases}$$

Then $\overline{\varphi}$ is one-to-one and onto by (III). We shall show that $\overline{\varphi}$ is order-preserving. Let $\lambda\leqq\mu$ in Λ_1. When $\lambda,\mu\in L$, we have $\overline{\varphi}(\lambda)=\lambda\leqq\mu=\overline{\varphi}(\mu)$. When $\lambda,\mu\in S_1$, we have $\overline{\varphi}(\lambda)=\varphi(\lambda)\leqq\varphi(\mu)=\overline{\varphi}(\mu)$. When $\lambda\in L$ and $\mu\in S_1$, we have $\lambda=0$ whence $\overline{\varphi}(\lambda)=0\leqq\overline{\varphi}(\mu)$. When $\lambda=u\in S_1$ and $\mu=a\in L$, then there is a point p of L with $p\leqq a$, since $a\neq 0$. By (1) we have $p\sqcup_2\varphi(u)=p\sqcup_1 u\leqq a$, whence $\overline{\varphi}(\lambda)=\varphi(u)\leqq a=\overline{\varphi}(\mu)$. Therefore $\overline{\varphi}$ is order-preserving and hence it is an isomorphism of Λ_1 onto Λ_2 which is the identity on L. □

Remark (21.10). If an atomistic Wilcox lattice L is of length 2, then L has two modular extensions $\Lambda_1\equiv L$ and $\Lambda_2\equiv L\cup\{i\}$.

Moreover, if a Wilcox lattice is not atomistic, its modular extension is not necessarily unique (see Exercise 21.1).

Later, in Section 25, we shall have a generalization of (21.9).

Theorem (21.11). *Let $L\equiv\Lambda-S$ be a Wilcox lattice where S is not empty. L is a matroid lattice if and only if its modular extension Λ is an irreducible matroid lattice and there exists the imaginary unit for L.*

Proof. The "if" part follows from (20.13.2). Conversely, let L be a matroid lattice. When L is of length $\geqq 4$, then by (20.9) and (20.6) L is a non-modular affine matroid lattice. Hence it follows from (19.14) and

(21.9) that Λ is an irreducible modular matroid lattice and there exists the imaginary unit for L.

When L is of length $\leqq 3$, it follows from (20.5) that Λ is a modular matroid lattice and that there is the imaginary unit for L. To prove that Λ is irreducible it suffices to show that if Λ is reducible then a Wilcox lattice $L \equiv \Lambda - S$ (S is not empty) is not atomistic. Let $\Lambda = \Lambda_1 \Lambda_2$ where Λ_i $(i=1,2)$ is complemented modular. Since the length of $\Lambda \leqq 3$, one of Λ_i, say Λ_2, is a trivial lattice $\{0_2, 1_2\}$. When the length of $\Lambda = 2$, we have $\Lambda_1 = \{0_1, 1_1\}$ and evidently L is not atomistic.

Let the length of $\Lambda = 3$ (the length of $\Lambda_1 = 2$). A point of Λ is either $[0_1, 1_2]$ or $[p_1, 0_2]$ for some point p_1 of Λ_1. When $[p_1, 1_2] \in S$ for some point $p_1 \in \Lambda_1$, since Λ_1 is complemented there is a point $q_1 \in \Lambda_1$ with $p_1 \neq q_1$, and then we have $[q_1, 1_2] \in L$, since otherwise

$$[1_1, 1_2] = [p_1, 1_2] \sqcup [q_1, 1_2] \in S.$$

Since $[0_1, 1_2] \in S$, the element $[q_1, 1_2]$ contains only one point $[q_1, 0_2]$ in L. Hence L is not atomistic. When $[p_1, 1_2] \in L$ for every point $p_1 \in \Lambda_1$, then

$$S = \Lambda_1 \times \{0_2\} - [0_1, 0_2] \quad \text{or} \quad S = \{[0_1, 1_2]\} \quad \text{or}$$
$$S = \{[p_1, 0_2]\} \quad \text{for some point } p_1 \in \Lambda_1.$$

In any case, L is not atomistic, since the element $[p_1, 1_2] \in L$ contains only one point in L. ▯

Lemma (21.12). *Let $L \equiv \Lambda - S$ be an atomistic Wilcox lattice of length $\geqq 3$. A line l of L is incomplete if and only if it is a singular element.*

Proof. This follows from (21.7.1). ▯

Theorem (21.13). *Let $L \equiv \Lambda - S$ be a complete atomistic Wilcox lattice of length $\geqq 3$ with the imaginary unit* (in particular, let L be a non-modular affine matroid lattice). *An element a of L with $h(a) \geqq 2$ is incomplete if and only if it is a singular element.*

Proof. If a is singular, then by (21.3.2) every line contained in a is singular. Hence a is incomplete by (21.12).

Conversely, let a be incomplete. By (8.16), a has a base $\{p, p_\alpha; \alpha \in I\}$ where I is not empty. For every $\alpha \in I$, since the line $l_\alpha = p \vee p_\alpha$ is singular by (21.12), we have $l_\alpha = p \sqcup \iota(l_\alpha)$ by (21.3.1). Since Λ is complete by (20.11) and since S has the greatest element, the join $\bigsqcup(\iota(l_\alpha); \alpha \in I)$ exists and belongs to S. Moreover

$$a = \bigvee_\alpha l_\alpha = \bigsqcup_\alpha l_\alpha = p \sqcup \bigsqcup_\alpha \iota(l_\alpha).$$

Hence a is singular. ▯

Theorem (21.14). (Modularity and parallelism). *Let* $L \equiv \Lambda - S$ *be an atomic Wilcox lattice and let a and b be two non-zero elements of L such that* $a \wedge b = 0$. *The pair* (a,b) *is modular if and only if there do not exist singular elements* a_1 *and* b_1 *such that*

$$a_1 \parallel b_1, \quad a_1 \leqq a \quad \textit{and} \ b_1 \leqq b.$$

Proof. (I) Assume $(a,b)M$. Since $a \wedge b = 0$, it follows from (3.11.6) that $a \sqcap b = 0$. If $a_1 \leqq a$ and $b_1 \leqq b$, then $a_1 \sqcap b_1 = 0 \notin S$. Hence by (20.7), $a_1 \parallel b_1$ does not hold.

(II) Assume $(a,b)\bar{M}$. We have $a \sqcap b \in S$ by (3.11.5). Since L is atomic there are two points p and q such that $p \leqq a$ and $q \leqq b$. Put $a_1 = p \sqcup (a \sqcap b)$ and $b_1 = q \sqcup (a \sqcap b)$. Then a_1 and b_1 are singular elements such that $a_1 \leqq a$ and $b_1 \leqq b$. Moreover, since $\iota(a_1) = \iota(b_1)$ and $a_1 \wedge b_1 = 0$, we have $a_1 \parallel b_1$ by (21.5.2). □

(We remark that if $a \wedge b \neq 0$ then $(a,b)M$ always since L is weakly modular.)

Corollary (21.15). *In an atomistic Wilcox lattice L, an element a with* $0 < a < 1$ *is modular if and only if a contains no singular element.*

Proof. If a contains no singular element, then it follows from (21.14) that $(b,a)M$ for every $b \in L$ with $a \wedge b = 0$. Since L is weakly modular, a is a modular element.

If a contains a singular element a_1, then taking a point p with $p \nleqq a$, the element $b = p \vee \iota(a_1)$ is singular and $b \parallel a_1$ by (21.5.2). Hence $(b,a)\bar{M}$ by (21.14). □

Corollary (21.16). *Let L be an affine matroid lattice.*

(21.16.1) *An element a of L with* $0 < a < 1$ *is modular if and only if a contains no incomplete line.*

(21.16.2) *If L satisfies Euclid's strong parallel axiom then any modular element a with* $0 < a < 1$ *is a point.*

Proof. (21.16.1) follows from (21.15) and (21.13). (21.16.2) follows from (21.16.1) and (18.10). □

Theorem (21.17). (Comparability theorem). *In a Wilcox lattice* $L \equiv \Lambda - S$, *we write* $\iota(p) = 0$ *when p is a point of L. Let a and b be two singular elements and let p and q be points of L contained in a and b respectively. Then there exist elements* a', a'', b' *and* b'' *such that*

$$a = a' \vee a'', \quad a' \wedge a'' = p, \quad b = b' \vee b'', \quad b' \wedge b'' = q,$$
$$a' \parallel\!| b', \quad \textit{and} \ \iota(a'') \sqcap \iota(b'') = 0.$$

Moreover $\iota(a') = \iota(b') = \iota(a) \sqcap \iota(b)$.

Proof. We put $u=\iota(a)\sqcap\iota(b)$ and we take complements v and w of u in $\Lambda[0,\iota(a)]$ and $\Lambda[0,\iota(b)]$ respectively. Then each of u, v and w is either zero or an element of S. Put $a'=p\sqcup u$, $a''=p\sqcup v$, $b'=q\sqcup u$ and $b''=q\sqcup w$. Then

$$a'\vee a''=a'\sqcup a''=p\sqcup u\sqcup v=p\sqcup\iota(a)=a,$$

and similarly $b'\vee b''=b$. Since $a'\sqcap a''\geqq p$, we have $a'\sqcap a''\in L$. By (3.11.4)

$$a'\wedge a''=a'\sqcap a''=(p\sqcup u)\sqcap(p\sqcup v).$$

Since $p\sqcap(u\sqcup v)=0$, the sublattice $\Lambda[0,u\sqcup v]$ is isomorphic to $\Lambda[p,p\sqcup u\sqcup v]$ by the mapping $\lambda\to p\sqcup\lambda$. Hence

$$(p\sqcup u)\sqcap(p\sqcup v)=p\sqcup(u\sqcap v)=p.$$

Therefore $a'\wedge a''=p$, and similarly $b'\wedge b''=q$.

We have $\iota(a')=\iota(b')=u=\iota(a)\sqcap\iota(b)$, and

$$\iota(a'')\sqcap\iota(b'')=v\sqcap w=v\sqcap\iota(a)\sqcap w\sqcap\iota(b)=u\sqcap v\sqcap w=0.$$

Moreover, when $u=0$, we have $a'\parallel\!\!\parallel b'$ since a' and b' are points, and when $u\neq 0$, we have $a'\parallel\!\!\parallel b'$ by (21.5.2). ▯

Remark (21.18). Let a be an incomplete element of an affine matroid lattice L. Since a is singular by (21.13), $a=p\sqcup\iota(a)$ for a point p of L with $p\leqq a$. Let q be a point of L. The $\parallel$-image $\varphi_q(a)$ of a at q, defined in (19.5), is an incomplete element such that $\varphi_q(a)\parallel\!\!\parallel a$ and $q<a$. Hence we have

(21.18.1) $\qquad \varphi_q(a)=q\sqcup\iota(a)$ for every point q of L.

We remark that (21.18.1) still holds when a is a point.

If each of a and b is either a point or an incomplete element, then by the same manner as in the proof of (21.17) we have

$$\varphi_q(a)\wedge\varphi_q(b)=(q\sqcup\iota(a))\sqcap(q\sqcup\iota(b))=q\sqcup(\iota(a)\sqcap\iota(b)).$$

Corollary (21.19). *In an affine matroid lattice L, let a and b be two incomplete elements and let p and q be points of L contained in a and b respectively. Then, there exist elements a', a'', b' and b'' such that*

$$a=a'\vee a'',\quad a'\wedge a''=p,\quad b=b'\vee b'',\quad b'\wedge b''=q,$$
$$a'\parallel\!\!\parallel b',\ \text{ and }\ \varphi_r(a'')\wedge\varphi_r(b'')=r\quad \text{for every point } r \text{ of } L.$$

Moreover, $\varphi_r(a')=\varphi_r(b')=\varphi_r(a)\wedge\varphi_r(b)$.

Proof. By the remarks in (21.18), this follows from (21.17). ▯

Corollary (21.20). *Let a and b be two incomplete elements of an affine matroid lattice L such that* $a \wedge b = 0$. *The following three statements are equivalent.*

(α) $(a,b)M$.
(β) *There do not exist incomplete elements* a_1 *and* b_1 *such that*

$$a_1 \parallel b_1, \quad a_1 \leqq a \quad \text{and } b_1 \leqq b.$$

(γ) $\varphi_r(a) \wedge \varphi_r(b) = r$ *for every point* $r \in L$.

Proof. The equivalence of (α) and (β) follows from (21.14).

By (21.18), (γ) is equivalent to $\imath(a) \sqcap \imath(b) = 0$. If $(a,b)M$, then since $a \wedge b = 0$, we have $a \sqcap b = 0$ by (3.11.6). Hence $\imath(a) \sqcap \imath(b) = 0$. If $(a,b)\bar{M}$, then since $a \sqcap b \in S$ by (3.11.5), we have $\imath(a) \geqq a \sqcap b$ and $\imath(b) \geqq a \sqcap b$. Hence $\imath(a) \sqcap \imath(b) \neq 0$. ▯

Exercise 21.1. Show that the third lattice in Exercise 1.1 has two modular extensions ($h(i) = 1$ in one of them and $h(i) = 2$ in the other).

22. Affine Matroid Lattices Satisfying Euclid's Strong Parallel Axiom

Lemma (22.1). *Let* $L \equiv \Lambda - S$ *be a Wilcox lattice. For any non-zero element* $a \in L$, *the sublattice* $L[0,a]$ *is a Wilcox lattice with a modular extension* $\Lambda[0,a]$. *If there exists the imaginary unit i for L and if* $a \sqcap i \neq 0$ *then* $a \sqcap i$ *is the imaginary unit for* $L[0,a]$.

Proof. Evidently $\Lambda[0,a]$ is a complemented modular lattice and the subset $S_a = S \cap \Lambda[0,a]$ of $\Lambda[0,a]$ satisfies the conditions (3.11.1) and (3.11.2). Hence $L[0,a] \equiv \Lambda[0,a] - S_a$ is a Wilcox lattice. If i is the greatest element of S and $a \sqcap i \neq 0$ then evidently $a \sqcap i$ is the greatest element of S_a. ▯

Theorem (22.2). *Let L be a non-modular affine matroid lattice and let a be an element of L with* $h(a) \geqq 4$. *Then,* $L[0,a]$ *is an affine matroid lattice, and* $L[0,a]$ *satisfies Euclid's strong parallel axiom if and only if a is an incomplete element of L.*

Proof. By (19.14), L is a Wilcox lattice with the modular extension Λ and with the imaginary unit i. By (22.1), $L[0,a]$ is a Wilcox lattice. Since L is compactly atomistic, so is $L[0,a]$. Hence it follows from (20.9) that $L[0,a]$ is an affine matroid lattice. When $a \sqcap i = 0$, then $L[0,a]$ is modular since $L[0,a] = \Lambda[0,a]$. When $a \sqcap i \neq 0$, by (22.1) $a \sqcap i$ is the imaginary unit for $L[0,a]$. By (20.15), $L[0,a]$ satisfies

Euclid's strong parallel axiom if and only if $a \sqcap i$ is a hyperplane of $\Lambda[0,a]$, that is,

(1) $$a \sqcap i \lessdot a \quad \text{in } \Lambda.$$

It is easy to show that (1) holds if and only if a is a singular element of L. By (21.13.2), (1) holds if and only if a is an incomplete element. □

Definition (22.3). Let L be a non-modular affine matroid lattice. It follows from (18.17) that there exists a maximal incomplete element $I(p)$ for each point p of L. We denote by Ω_I the set $\{I(p); p \in \Omega(L)\}$. Two different elements of Ω_I are always parallel by (18.17). By (21.13), Ω_I is equal to the set of maximal singular elements $p \sqcup i$, where i is the imaginary unit for L.

If an affine matroid lattice L is modular, then by (18.20) we may put $\Omega_I = \Omega(L)$ and then Ω_I forms a projective space.

Definition (22.4). Let L be a non-modular affine matroid lattice. We denote by $M(L)$ the set of all elements a of L satisfying the following condition:

(22.4.1) If p is a point contained in a then every incomplete line which contains p is contained in a.

By (18.18), this condition is equivalent to the following one:

(22.4.2) If p is a point contained in a then $I(p) \leqq a$.

Evidently, $0, 1 \in M(L)$ and $I(p) \in M(L)$ for every $p \in \Omega(L)$. $M(L)$ is called the *modular center* of L.

If an affine matroid lattice L is modular, then we may put $M(L) = L$.

Lemma (22.5). *A non-zero element a of a non-modular affine matroid lattice L belongs to $M(L)$ if and only if $a > i$ where i is the imaginary unit for L.*

Proof. Since $I(p) = p \sqcup i$ in the modular extension of L, $i < a$ if and only if $I(p) \leqq a$ for any $p \leqq a$. □

Theorem (22.6). *Let L be a non-modular affine matroid lattice. The set Ω_I of all maximal incomplete elements of L forms an irreducible projective space, and the modular center $M(L)$ is an irreducible modular matroid lattice which is isomorphic to the projective geometry on Ω_I.*

Proof. (I) We define a mapping φ of $M(L)$ into Λ as follows:

$$\varphi(a) = \begin{cases} a & \text{if } a \neq 0, \\ i & \text{if } a = 0. \end{cases}$$

By (22.5), φ is one-to-one and order-preserving and moreover the image of φ is equal to the sublattice $\Lambda[i,1]$. Hence $M(L)$ forms a modular matroid lattice isomorphic to $\Lambda[i,1]$. The point space of $M(L)$ is the set $\{p \sqcup i; p \in \Omega(L)\}$ which is equal to Ω_I. Hence it follows from (16.5) that Ω_I forms a projective space and that $M(L)$ is isomorphic to the lattice of linear subsets of Ω_I.

(II) We shall show that Ω_I is irreducible, that is, every line of Ω_I contains at least three points (see (16.7)). If $I(p)$ and $I(q)$ are different points of Ω_I, then the line $p \vee q$ of L is complete, since otherwise $p \vee q \leqq I(p)$, whence $I(p) = I(q)$. By (18.7), $p \vee q$ contains a third point r. Since $p \vee r = q \vee r = p \vee q$ is a complete line, we have $I(r) \neq I(p)$ and $I(r) \neq I(q)$. Moreover $I(r) = r \sqcup i \leqq p \sqcup q \sqcup i = I(p) \vee I(q)$. Hence Ω_I is irreducible. By (16.7), $M(L)$ is irreducible. □

Remark (22.7). We compare the results (22.6) and (22.2) with the central decomposition of a matroid lattice (see (10.12), (10.13) and (13.6)).

Let L be a non-modular affine matroid lattice:	Let L be a matroid lattice:
(1) The point space Ω_I of the modular center $M(L)$ is an irreducible projective space.	(1) The point space Ω_0 of the center $Z(L)$ is a set.
(2) The modular center $M(L)$ is an irreducible modular matroid lattice isomorphic to the lattice of all linear subsets of Ω_I.	(2) The center $Z(L)$ is a complete Boolean atomistic lattice (= a distributive matroid lattice) isomorphic to the lattice of all subsets of Ω_0.
(3) For any element $I(p) \in \Omega_I$, the sublattice $L[0, I(p)]$ satisfies Euclid's strong parallel axiom.	(3) For any element $z \in \Omega_0$, the sublattice $L[0,z]$ is irreducible.
(4) Ω_I is a singleton set if and only if for any different points p and q of L there exists $a \in L$ such that $$p \vee q \parallel a.$$	(4) Ω_0 is a singleton set if and only if for any different points p and q of L there exists $a \in L$ such that $$p \sim_a q.$$

For a non-modular affine matroid lattice L, by the above consideration, we may call Ω_I the decomposition space of the parallel decomposition of L and we may say that L is *modularly irreducible* when L satisfies Euclid's strong parallel axiom.

Definition (22.8). Let E be a vector space over a division ring K. A subset A of E is called an *affine subset* when

(22.8.1) $x_1,\ldots,x_n\in A$, $\lambda_1,\ldots,\lambda_n\in K$ and $\lambda_1+\cdots+\lambda_n=1$

imply $\lambda_1 x_1+\cdots+\lambda_n x_n\in A$.

The empty set $\emptyset$ and any singleton set $\{x\}$ are affine subsets. The set of all affine subsets of E is denoted by $L_A(E)$.

Theorem (22.9). *Let E be a vector space over a division ring K. The set $L_A(E)$ of all affine subsets of E forms a compactly atomistic Wilcox lattice, ordered by set-inclusion. If the dimension of E is not less than 3, then $L_A(E)$ is an affine matroid lattice satisfying Euclid's strong parallel axiom.*

Proof. (I) It is easy to verify that $L_A(E)$ satisfies the three conditions in (15.3). Hence $L_A(E)$ forms an upper continuous lattice. $L_A(E)$ is atomistic since any singleton set of E is a point of $L_A(E)$. Hence $L_A(E)$ is a compactly atomistic lattice.

(II) Let $\hat{E}$ be the product space of E with a one-dimensional vector space $E_1=Kx_0$, and let $L(\hat{E})$ be the set of all subspaces of $\hat{E}$. By (16.9), $L(\hat{E})$ forms an irreducible modular matroid lattice ordered by set-inclusion, and $\hat{H}=E\times\{0\}$ is a hyperplane of $L(\hat{E})$. We define a mapping φ of $L_A(E)$ into $L(\hat{E})$ as follows: $\varphi(\emptyset)=\{0\}$, and for a non-empty $A\in L_A(E)$, $\varphi(A)$ is the subspace of $\hat{E}$ generated by the set $\{(x,x_0);x\in A\}$. Evidently φ is order-preserving. Since A is an affine subset, it is easy to verify that

$$A=\{x\in E;(x,x_0)\in\varphi(A)\}.$$

Hence φ is one-to-one.

(III) We shall show that a non-zero dimensional subspace $\hat{M}$ of $\hat{E}$ belongs to the image of φ if and only if $\hat{M}\nleq\hat{H}$ in $L(\hat{E})$. It is evident that $\varphi(A)\nleq\hat{H}$ for every non-empty $A\in L_A(E)$. Conversely, assume $\hat{M}\nleq\hat{H}$. We put

$$A=\{x\in E;(x,x_0)\in\hat{M}\}.$$

It is easy to show that A is an affine subset of E. A is not empty, because if we take an element $(x,\lambda x_0)\in\hat{M}$ such that $(x,\lambda x_0)\notin\hat{H}$, then since $\lambda\neq 0$, we have $(\lambda^{-1}x,x_0)\in\hat{M}$, whence $\lambda^{-1}x\in A$. Evidently $\varphi(A)\leq\hat{M}$. We shall show $\varphi(A)\geq\hat{M}$. If $(x,\lambda x_0)\in\hat{M}$, then since $(\lambda^{-1}x,x_0)\in\hat{M}$, we have $\lambda^{-1}x\in A$, whence $(x,\lambda x_0)=\lambda(\lambda^{-1}x,x_0)\in\varphi(A)$. If $(x,0)\in\hat{M}$, then we take an element $y\in A$. Since $(y,x_0)\in\hat{M}$, we have $(x+y,x_0)\in\hat{M}$, whence $x+y\in A$. Hence $(x,0)=(x+y,x_0)-(y,x_0)\in\varphi(A)$. Therefore $\varphi(A)=\hat{M}$.

(IV) Let $S=\{\hat{M}\in L(\hat{E});\{0\}\neq\hat{M}\leq\hat{H}\}$. By (II) and (III), φ is an order-preserving, one-to-one mapping of $L_A(E)$ onto the Wilcox lattice $L\equiv L(\hat{E})-S$. Hence $L_A(E)$ is isomorphic to L.

(V) If the dimension of E is not less than 3, then the length of $L_A(E)$ is not less than 4. Hence $L_A(E)$ is an affine matroid lattice by (20.9). Since the hyperplane $\hat{H}$ of $L(\hat{E})$ is the imaginary unit for L, by (20.15) L satisfies Euclid's strong parallel axiom, and so does $L_A(E)$. ▯

Remark (22.10). A set Ω, whose elements are called points, is called an *affine space* when there exists a family of subsets of Ω, called lines, and there exists a family of subsets of Ω, called planes, satisfying the following four conditions (cf. Sasaki [1]):

(AS 1) Every line contains at least two points and any two different points determine a line.

(AS 2) Every plane contains three non-collinear points and any three non-collinear points determine a plane.

By a subspace of Ω we mean a subset S such that if p and q are different points of S then the line determined by p and q is contained in S and if p, q and r are non-collinear points of S then the plane determined by them is contained in S. By a three-space of Ω we mean the least subspace containing four points which are not in a plane.

(AS 3) There exists a three-space, and any two planes in a three-space cannot have just one point in common.

(AS 4) If l is a line then for any point p which is not contained in l there exists one and only one line k such that k contains p, l and k are disjoint, and l and k are contained in the same plane.

It is easy to show that the point space $\Omega(L)$ of an affine matroid lattice L satisfying Euclid's strong parallel axiom forms an affine space. Conversely, it is proved in Sasaki [1] and [3] that the set $L(\Omega)$ of all subspaces of an affine space Ω forms an affine matroid lattice satisfying Euclid's strong parallel axiom.

References for Chapter IV

For Section 17: M. L. Dubreil-Jacotin, L. Leisieur and R. Croisot [1] (Part III), C.-J. Hsu [1], F. Maeda [5].

For Section 18: M. L. Dubreil-Jacotin, L. Leisieur and R. Croisot [1] (Part III), F. Maeda [4].

For Section 19 and 20: M. L. Dubreil-Jacotin, L. Leisieur and R. Croisot [1] (Part III), F. Maeda [5].

For Section 21: F. Maeda [5].

For Section 22: F. Maeda [4] and [6], U. Sasaki [1] and [3].

Chapter V

Point-free Parallelism in Symmetric Lattices

23. Point-free Parallelism in Lattices

Definition (23.1). Let L be a lattice with 0 and let a and b be non-zero elements of L. When

(23.1.1) $a \wedge b = 0$ and there exists a modular element m such that $m \vee b = a \vee b$ and $m \leqq a$,

then we write $a \underset{(m)}{<\!|}\, b$. We have $0 < m < 1$, since if $m = 0$ then $b = a \vee b$, whence $a = a \wedge b = 0$, and if $m = 1$ then $a = 1$, whence $b = a \wedge b = 0$.

When $a \underset{(m)}{<\!|}\, b$ and $b \underset{(n)}{<\!|}\, a$, we say that a and b are *parallel* with axes m and n, and we write $a \underset{(m,n)}{\parallel} b$. Since modular elements m and n are not necessarily points, we may say that this parallelism is *point-free*.

Remark (23.2). It follows from (17.13) that, in an atomic lattice with the covering property, $a <\!| \, b$ is equivalent to $a \underset{(p)}{<\!|}\, b$ for some point $p \leqq a$, and $a \parallel b$ is equivalent to $a \underset{(p,q)}{\parallel} b$ for points $p \leqq a$ and $q \leqq b$. Hence, if an affine matroid lattice L satisfies Euclid's strong parallel axiom then by (21.16.2) the point-free parallelism in L coincides with the usual parallelism. But, if L does not satisfy the axiom then any two complete lines l and k with $l \wedge k = 0$ are point-free parallel, since any complete line is a modular element by (21.16.1).

Lemma (23.3) *Let* $a \underset{(m)}{<\!|}\, b$ *in a lattice* L *with* 0.

(23.3.1) *If* $m < a_1 \leqq a$ *then* $(b, a_1)M$ *does not hold.*

(23.3.2) *If* $m \leqq a_1 < a$ *then* $(b, a_1)M^*$ *does not hold.*

(23.3.3) *If* L *is* M*-symmetric and if* $m \leqq a_1 \leqq a$ *then* $(a_1, b)M^*$ *holds.*

Proof. (I) If $m < a_1 \leqq a$, then since $m \vee b = a_1 \vee b = a \vee b$ and $a_1 \wedge b = 0$, we have

$$(m \vee b) \wedge a_1 = (a_1 \vee b) \wedge a_1 = a_1 > m = m \vee (b \wedge a_1).$$

Hence $(b,a_1)M$ does not hold.

(II) If $m \leqq a_1 < a$, then

$$(a \wedge b) \vee a_1 = a_1 < a = a \wedge (b \vee a) = a \wedge (b \vee a_1).$$

Hence $(b,a_1)M^*$ does not hold.

(III) Assume that L is M-symmetric, and let $m \leqq a_1 \leqq a$. Let $c \geqq b$. Since m is modular, by the assumption we have $(m,c)M$. Hence

$$(c \wedge a_1) \vee b \geqq b \vee (m \wedge c) = (b \vee m) \wedge c = c \wedge (a_1 \vee b) \geqq (c \wedge a_1) \vee b.$$

Therefore $(a_1,b)M^*$ holds. □

Remark (23.4). If a lattice L with 0 is M-symmetric and M*-symmetric, then it follows from (23.3.2) and (23.3.3) that $a \underset{(m)}{<|} b$ implies $a=m$, and hence L has only trivial parallelism.

Lemma (23.5). *Let a and b be non-zero elements of a lattice L with* 0. *Then,* $a \underset{(m,n)}{\parallel} b$ *if and only if*

(23.5.1) $a \wedge b = 0$ *and there exist modular elements m and n such that* $m \leqq a$, $n \leqq b$ *and* $a \vee n = b \vee m$.

Proof. If $a \underset{(m,n)}{\parallel} b$ then we have $b \vee m = a \vee b = a \vee n$. Hence (23.5.1) holds. If (23.5.1) holds then since $a \leqq m \vee b$, we have $a \vee b \leqq m \vee b \leqq a \vee b$, whence $a \underset{(m)}{<|} b$. Similarly we have $b \underset{(n)}{<|} a$. □

Lemma (23.6). *In a lattice L with* 0, *if* $a \underset{(m)}{<|} b$ *and* $m \leqq a_1 < a$, *then* $a_1 \underset{(m)}{<|} b$.

Proof. Evidently $a \wedge b = 0$ implies $a_1 \wedge b = 0$ and $m \vee b = a \vee b$ implies $m \vee b = a_1 \vee b$. Hence $a_1 \underset{(m)}{<|} b$. □

Lemma (23.7). *In a lattice L with* 0, *let* $a \underset{(m)}{<|} b$ *and* $b < b_2$. *If* $a \wedge b_2 = 0$ *then* $a \underset{(m)}{<|} b_2$, *and if* $m < b_2$ *then* $a < b_2$.

Proof. Since $m \vee b = a \vee b$ and $b < b_2$, we have $m \vee b_2 = a \vee b_2$. Hence, if $a \wedge b_2 = 0$ then $a \underset{(m)}{<|} b_2$ and if $m < b_2$ then $b_2 = a \vee b_2 \geqq a \vee b > a$. □

Lemma (23.8). *In a weakly modular lattice L, if* $a \underset{(m)}{<|} b$ *and if n is a non-zero modular element with* $n \leqq b$, *then* $a \underset{(m,n)}{\parallel} (a \vee n) \wedge b$.

Proof. Put $b_1 = (a \vee n) \wedge b$. By the weak modularity of L we have $(b, a \vee n)M$, since $b \wedge (a \vee n) \geqq n > 0$.

Since $m \leqq a \leqq a \vee n$, we have

$$b_1 \vee m = m \vee \{b \wedge (a \vee n)\} = (m \vee b) \wedge (a \vee n)$$
$$= (a \vee b) \wedge (a \vee n) = a \vee n.$$

Since $b_1 \geqq n$ and $a \wedge b_1 \leqq a \wedge b = 0$, we have $a \underset{(m,n)}{\parallel} b_1$ by (23.5). □

Theorem (23.9). (Parallel mappings) *Let $a \underset{(m,n)}{\parallel} b$ in a weakly modular M-symmetric lattice L. Put*

$$\varphi(a_1) = (a_1 \vee n) \wedge b \quad \textit{for } a_1 \in L[m,a] \quad \textit{and}$$
$$\psi(b_1) = (b_1 \vee m) \wedge a \quad \textit{for } b_1 \in L[n,b].$$

Then φ and ψ are mutually inverse, isomorphic mappings between the lattices $L[m,a]$ and $L[n,b]$.

Two elements $a_1 \in L[m,a]$ and $b_1 \in L[n,b]$ correspond by these mappings if and only if the following equality holds:

(1) $$a_1 \vee n = b_1 \vee m.$$

In this case $a_1 \underset{(m,n)}{\parallel} b_1$.

Proof. (I) It is evident that $\varphi(a_1) \in L[n,b]$ and $\psi(b_1) \in L[m,a]$. Since $a_1 \underset{(m)}{<\!|} b$ by (23.6), it follows from (23.8) that $a_1 \underset{(m,n)}{\parallel} \varphi(a_1)$. Similarly we have $b_1 \underset{(n,m)}{\parallel} \psi(b_1)$. By (23.5), (1) holds when either $\varphi(a_1) = b_1$ or $\psi(b_1) = a_1$.

(II) Conversely, assume that (1) holds. Since L is M-symmetric, we have $(m,b)M$. Hence

$$\varphi(a_1) = (a_1 \vee n) \wedge b = (b_1 \vee m) \wedge b = b_1 \vee (m \wedge b) = b_1.$$

Similarly $\psi(b_1) = a_1$.

(III) For any $a_1 \in L[m,a]$, if we put $b_1 = \varphi(a_1)$, then since (1) holds by (I), we have $\psi(b_1) = a_1$ by (II). Hence $\psi\varphi(a_1) = a_1$. Similarly $\varphi\psi(b_1) = b_1$ for every $b_1 \in L[n,b]$. Therefore φ and ψ are mutually inverse, one-to-one mappings between $L[m,a]$ and $L[n,b]$. Since they preserve the order, they are isomorphic mapping. □

Exercise 23.1. Let $a \underset{(m)}{<\!|} b$ in a lattice with 0 and assume that a is not modular. Prove that $b \triangledown a$ does not hold.

Exercise 23.2. Prove that if $a \underset{(m)}{<\!|} b$ in a lattice with 0 then m is a maximal modular element contained in a.

Exercise 23.3. In a non-modular affine matroid lattice which does not satisfy Euclid's strong parallel axiom, find a pair of non-zero elements a and b and an element b_2 with $b < b_2$ satisfying the following conditions:

$$a \underset{(m)}{<\!|} b \quad \text{but neither} \quad a \underset{(m)}{<\!|} b_2 \quad \text{nor} \quad a < b_2.$$

24. Point-free Parallelism in Wilcox Lattices

Theorem (24.1). (Parallelism and modularity) *In a Wilcox lattice* $L \equiv \Lambda - S$, *let a and b be non-zero elements of L and let m be a modular element with* $0 < m < a$. *Then* $a \underset{(m)}{<|} b$ *in L if and only if*

(24.1.1) $(b, a_1)\bar{M}$ *for every element* $a_1 \in L$ *such that* $m < a_1 \leqq a$.

Proof. If $a \underset{(m)}{<|} b$ then (24.1.1) holds by (23.3.1). Conversely, assume that (24.1.1) holds. Since L is weakly modular and since $(b,a)\bar{M}$, we have $a \wedge b = 0$. Since $b \wedge m \leqq a \wedge b = 0$ and since $(b,m)M$, we have $b \sqcap m = 0$ by (3.11.6). Putting $a_1 = m \sqcup (a \sqcap b)$, we have $m \leqq a_1 \leqq a$. Let a_2 be a complement of a_1 in $\Lambda[m,a]$. Evidently $a_1, a_2 \in L$. By the modularity of Λ,

$$a_1 \sqcap b = \{(a \sqcap b) \sqcup m\} \sqcap b = (a \sqcap b) \sqcup (m \sqcap b) = a \sqcap b,$$

whence $a_2 \sqcap b = a_2 \sqcap a \sqcap b = a_2 \sqcap a_1 \sqcap b = m \sqcap b = 0$. Hence $(b, a_2)M$ by (3.11.5). We have $a_2 = m$, since otherwise $(b, a_2)\bar{M}$ by the assumption. Hence $a_1 = a$. Since $m \vee b \geqq a_1 = a$, we have $m \vee b = a \vee b$. Therefore $a \underset{(m)}{<|} b$. □

Corollary (24.2). *In an atomistic Wilcox lattice* $L \equiv \Lambda - S$, *let a and b be non-zero elements. Then* $a <| b$ *if and only if* $(b, a_1)\bar{M}$ *for every element* a_1 *such that* $a_1 \leqq a$ *and* $h(a_1) \geqq 2$.

Proof. This follows from (24.1), since $a <| b$ is equivalent to $a \underset{(p)}{<|} b$ for some point $p \leqq a$ by (23.2). □

Definition (24.3). Let $L \equiv \Lambda - S$ be a Wilcox lattice. A non-zero element a of L is called a *regular element* when $a \sqcap u = 0$ for every $u \in S$. The set of all regular elements of L is denoted by R.

If $a \in R$ and $0 < a_1 \leqq a$ in L then $a_1 \in R$ evidently, and if L has a point p then $p \in R$ by (20.2.3). If S is empty then $R = L - \{0\}$, and if S is not empty then $1 \notin R$.

Lemma (24.4). *Let* $L \equiv \Lambda - S$ *be a Wilcox lattice. Any regular element of L is modular. Conversely, any modular element m of L with* $0 < m < 1$ *is regular if L is semicomplemented* (see (4.17)).

Proof. (I) If a is a regular element of L, then for every $b \in L$ we have $a \sqcap b \in L$, since otherwise $u = a \sqcap b$ is an element of S such that $a \sqcap u \neq 0$. Hence we have $(b, a)M$ by (3.11.5).

(II) Assume that L is semicomplemented. Let m be a modular element of L with $0 < m < 1$. Then there exists a non-zero element a of L such that $m \wedge a = 0$. Since $(a, m)M$, we have $m \perp a$, whence $m \sqcap a = 0$

by (3.11.6). For an arbitrary element u of S, we put $b=(m\sqcap u)\sqcup a$. Since $b\in L$, we have $(b,m)M$, whence $b\sqcap m\in L$ by (3.11.5). Moreover, by the modularity of Λ, we have

$$b\sqcap m=\{(m\sqcap u)\sqcup a\}\sqcap m=(m\sqcap u)\sqcup(a\sqcap m)=m\sqcap u.$$

Hence $m\sqcap u\in L$, which implies $m\sqcap u=0$. Therefore m is regular. □

Remark (24.5). Henceforth, in a Wilcox lattice $L\equiv\Lambda-S$, we use only regular elements as axes of point-free parallelism.

Definition (24.6). Let $L\equiv\Lambda-S$ be a Wilcox lattice. An element a of L is called an *irregular element* when there exists a regular element m of L and an element u of S such that $a=m\sqcup u$. The element u is called an *imaginary part* of a and we write $u=\imath(a)$. When a is a regular element, we may put $\imath(a)=0$.

Remark (24.7). Assume that a Wilcox lattice $L\equiv\Lambda-S$ has the imaginary unit i.

(24.7.1) A non-zero element a of L is regular if and only if $a\sqcap i=0$.

(24.7.2) Any non-zero element of L contains a regular element.

(24.7.3) A non-zero element of L is either regular or irregular.

Proof. (24.7.1) is evident. If $a\in L$ and $a\neq 0$, then taking a complement m of $a\sqcap i$ in $\Lambda[0,a]$, we have

$$m\neq 0,\qquad m\in L\quad\text{and } m\sqcap i=m\sqcap a\sqcap i=0.$$

Hence m is a regular element contained in a. Moreover, if $a\sqcap i\neq 0$, then a is irregular since $a=m\sqcup(a\sqcap i)$ and $a\sqcap i\in S$. □

Lemma (24.8). *Let a be an irregular element of a Wilcox lattice $L\equiv\Lambda-S$. An imaginary part $\imath(a)$ is uniquely determined by a, and it is the greatest element of S contained in a.*

Proof. Let $a=m\sqcup u$, $m\in R$ and $u\in S$. If $v\in S$ and $v\leqq a$, then since $u\sqcup v\in S$, we have $m\sqcap(u\sqcup v)=0$. Hence by the modularity of Λ we have

$$u=u\sqcup\{m\sqcap(u\sqcup v)\}=(u\sqcup m)\sqcap(u\sqcup v)=a\sqcap(u\sqcup v)\geqq v.$$

Therefore u is the greatest element of S contained in a, and hence $\imath(a)$ is uniquely determined by a. □

Remark (24.9). A Wilcox lattice $L\equiv\Lambda-S$ has the imaginary unit i if and only if 1 is an irregular element. For, if 1 is an irregular element then $i=\imath(1)$ is the greatest element of S by (24.8), and conversely, if there exists the imaginary unit i then since $1\notin R$, it follows from (24.7.3) that 1 is irregular.

Lemma (24.10). *Let a and b be non-zero elements of a Wilcox lattice $L \equiv \Lambda - S$. The following two statements are equivalent.*

(α) $a \underset{(m)}{<\!|} b$ *with* $a \neq m \in R$.

(β) $a \sqcap b \in S$ *and* $a = m \sqcup (a \sqcap b)$ *with* $m \in R$.

Each of (α) *and* (β) *implies that a is irregular and* $\iota(a) = a \sqcap b$.

Proof. (α)$\Rightarrow$(β). Since $(b, a)\bar{M}$ in L by (23.3.1), we have $a \sqcap b \in S$ by (3.11.5). Since $m \sqcup b = m \vee b = a \vee b = a \sqcup b$, by the modularity of Λ we have

$$m \sqcup (b \sqcap a) = (m \sqcup b) \sqcap a = (a \sqcup b) \sqcap a = a.$$

(β)$\Rightarrow$(α). By (3.11.4), $a \sqcap b \in S$ implies $a \wedge b = 0$. Since $a \sqcup b = m \sqcup (a \sqcap b) \sqcup b = m \sqcup b$, we have $a \vee b = m \vee b$. Hence $a \underset{(m)}{<\!|} b$. Moreover $m \neq a$ since $m \sqcap (a \sqcap b) = 0$ and $a \sqcap b \neq 0$.

Evidently (β) implies that a is irregular and $\iota(a) = a \sqcap b$. □

Lemma (24.11). *Let a and b be irregular elements of a Wilcox lattice $L \equiv \Lambda - S$. If $a \underset{(m)}{<\!|} b$ with $m \in R$ then $\iota(a) \leqq \iota(b)$, and if $a \underset{(m,n)}{\parallel} b$ with $m, n \in R$ then $\iota(a) = \iota(b)$.*

Proof. Let $a \underset{(m)}{<\!|} b$ with $m \in R$. Since a is irregular, we have $a \neq m$. Hence $\iota(a) = a \sqcap b \leqq b$ by (24.10). Therefore $\iota(a) \leqq \iota(b)$ by (24.8). □

Lemma (24.12). *In a Wilcox lattice $L \equiv \Lambda - S$, let $a = m \sqcup u$ and $b = n \sqcup u$ with $m, n \in R$ and $u \in S$. If either $a \wedge n = 0$ or $b \wedge m = 0$ then $a \underset{(m,n)}{\parallel} b$.*

Proof. Assume $a \wedge n = 0$. Since n is modular, we have $a \perp n$, whence $a \sqcap n = 0$. Hence

$$a \sqcap b = (u \sqcup n) \sqcap a = u \sqcup (n \sqcap a) = u \in S.$$

By (24.10) we have $a \underset{(m)}{<\!|} b$ and $b \underset{(n)}{<\!|} a$, whence $a \underset{(m,n)}{\parallel} b$. By symmetry, $b \wedge m = 0$ implies $a \underset{(m,n)}{\parallel} b$. □

Theorem (24.13). *In Wilcox lattice $L \equiv \Lambda - S$, if a is an irregular element such that $a = m \sqcup \iota(a)$ and $m \in R$, then for any regular element n with $a \wedge n = 0$ there exists one and only one element b such that $a \underset{(m,n)}{\parallel} b$. In this case, b is irregular and $b = n \sqcup \iota(a)$.*

Proof. Put $b = n \sqcup \iota(a)$. It follows from (24.12) that $a \underset{(m,n)}{\parallel} b$. If there exists b' such that $a \underset{(m,n)}{\parallel} b'$, then by (24.10) we have $a \sqcap b' \in S$ and $a = m \sqcup (a \sqcap b')$. Hence $a \sqcap b' = \iota(a)$. Since $n < n \sqcup \iota(a) \leqq b'$, by (24.10) again we have

$$b' = n \sqcup (a \sqcap b') = n \sqcup \iota(a) = b. \quad \square$$

Corollary (24.14). *In a Wilcox lattice $L \equiv \Lambda - S$, if $a \underset{(m)}{<\!|} b$ with $a \neq m \in R$, then for any regular element n with $n \leqq b$ there exists one and only one element b_1 such that $a \underset{(m,n)}{\parallel} b_1$. In this case, b_1 is irregular and $b_1 = n \sqcup (a \sqcap b) \leqq b$.*

Proof. By (24.10) we have $a \sqcap b \in S$ and $a = m \sqcup (a \sqcap b)$. Since $a \wedge n \leqq a \wedge b = 0$, it follows from (24.13) that there exists one and only one element b_1 such that $a \underset{(m,n)}{\parallel} b_1$, and moreover $b_1 = n \sqcup (a \sqcap b) \leqq b$. □

Theorem (24.15). (Modularity and parallelism) *Let $L \equiv \Lambda - S$ be a Wilcox lattice, and let a and b be two non-zero elements of L such that each of a and b contains a regular element and $a \wedge b = 0$. The pair (a,b) is modular if and only if there do not exist irregular elements a_1 and b_1 such that*

$$a_1 \leqq a, \quad b_1 \leqq b \quad \text{and } a_1 \underset{(m,n)}{\parallel} b_1 \quad \text{with } m,n \in R.$$

Proof. (I) Assume $(a,b)M$. Since $a \wedge b = 0$, we have $a \perp b$, whence $a \sqcap b = 0$. Then, for any $a_1 \leqq a$ and $b_1 \leqq b$, we have $a_1 \sqcap b_1 = 0 \notin S$. Hence $a_1 \underset{(m,n)}{\parallel} b_1$ does not hold by (24.10).

(II) Assume $(a,b)\overline{M}$. We have $a \sqcap b \in S$ by (3.11.5). Since a and b contain regular elements m and n respectively, putting $a_1 = m \sqcup (a \sqcap b)$ and $b_1 = n \sqcup (a \sqcap b)$, we have $a_1 \leqq a$ and $b_1 \leqq b$. Since $a_1 \wedge b_1 \leqq a \wedge b = 0$, we have $a_1 \underset{(m,n)}{\parallel} b_1$ by (24.12). □

Theorem (24.16). (Comparability theorem). *Let $L \equiv \Lambda - S$ be a Wilcox lattice, and let a and b be two irregular elements of L such that $a \wedge b = 0$. Then there exist $a', a'', b', b'' \in L$ and $m,n \in R$ such that*

$$a = a' \vee a'', \qquad a' \wedge a'' = m,$$

$$b = b' \vee b'', \qquad b' \wedge b'' = n,$$

$$a' \underset{(m,n)}{\parallel} b' \quad \text{and } \iota(a'') \sqcap \iota(b'') = 0.$$

Moreover $\iota(a') = \iota(b') = \iota(a) \sqcap \iota(b)$.

Proof. Let $a = m \sqcup \iota(a)$ and $b = n \sqcup \iota(a)$ with $m,n \in R$. We put $u = \iota(a) \sqcap \iota(b)$ and take complements v and w of u in $\Lambda[0, \iota(a)]$ and $\Lambda[0, \iota(b)]$ respectively. Then $u, v, w \in S \cup \{0\}$. Put $a' = m \sqcup u$, $a'' = m \sqcup v$, $b' = n \sqcup u$ and $b'' = n \sqcup w$. Since $a' \wedge b' \leqq a \wedge b = 0$, it follows from (24.12) that $a' \underset{(m,n)}{\parallel} b'$ when $u \neq 0$. When $u = 0$, evidently $a' \underset{(m,n)}{\parallel} b'$. The other equations are easily verified as in the proof of (21.17). □

Exercise 24.1. Find a Wilcox lattice having a modular element m with $0 < m < 1$ which is not regular.

25. Uniqueness of the Modular Extension of a Wilcox Lattice

First we give a generalization of (20.6).

Lemma (25.1). *Let $L \equiv \Lambda - S$ be a semicomplemented Wilcox lattice and assume that there exist two elements $a, b \in L$ such that $0 < a < b < 1$. Then, L is modular if and only if S is empty.*

Proof. If S is empty then evidently L is modular. Assume that S is not empty. First we shall prove that

(1) if there exist $u \in S$ and $a \in L$ such that $u < a < 1$ then L is not modular.

Since L is semicomplemented, there exists $c \in L$ such that $c \neq 0$ and $c \wedge a = 0$. Then by (3.11.4), we have $c \sqcap a \in S \cup \{0\}$. Putting $b = u \sqcup c$, we have $b \in L$ and

$$b \sqcap a = (u \sqcup c) \sqcap a = u \sqcup (c \sqcap a) \in S.$$

Hence $(b, a)\overline{M}$ in L by (3.11.5). Thus (1) has been proved.

When S is a singleton $\{u\}$, then evidently $0 \lessdot u$ in Λ. If $u \lessdot 1$ in Λ, then Λ would be an atomistic modular lattice of length 2, which contradicts the assumption that there exist $a, b \in L$ such that $0 < a < b < 1$. Hence there exists $\lambda \in \Lambda$ such that $u < \lambda < 1$. Evidently $\lambda \notin S$. Hence L is not modular by (1).

When S contains two elements, then there exist $u, v \in S$ such that $u < v$, since $S \cup \{0\}$ is an ideal. Let c be a complement of v in Λ. Then we have $0 \neq c \in L$, since $v \in S$ and $v \sqcup c = 1 \in L$. Hence $u \sqcup c \in L$. Since $\Lambda[0, v]$ is isomorphic to $\Lambda[c, 1]$ by the mapping $\lambda \to \lambda \sqcup c$, we have $u \sqcup c < v \sqcup c = 1$. Hence L is not modular by (1). ▯

Remark (25.2). In the above lemma, the assumption that there exist $a, b \in L$ such that $0 < a < b < 1$ is removable if L is not atomistic, because if this assumption does not hold then either 1 is an atom or every $a \in L$ with $0 < a < 1$ is an atom.

Let L be a lattice in the above lemma. It follows from (24.4) and (25.1) that regular elements of L can be defined without using imaginary elements. We may say that a non-zero element $a \in L$ with $a \neq 1$ is regular when a is a modular element, and that 1 is regular when L is a modular lattice.

Definition (25.3). Let L be a lattice with 0. We consider the following condition (MC) on L which is stronger than SSC.

(MC) If $a > b$ in L then there exists a modular element m of L such that $0 < m \leqq a$ and $m \wedge b = 0$.

Any atomistic lattice satisfies (MC), since every point is a modular element.

Remark (25.4). It is easy to show that a lattice L with 0 satisfies (MC) if and only if L is SSC and satisfies the following condition:

(25.4.1) Any non-zero element contains a non-zero modular element.

A Wilcox lattice $L \equiv \Lambda - S$ having the imaginary unit satisfies (MC) if it is SSC, since it satisfies (25.4.1) by (24.7.2) and (24.4).

Definition (25.5). Let $L \equiv \Lambda - S$ be a Wilcox lattice. When an irregular element a has the form $a = m \sqcup \imath(a)$ with $m \in R$, we call m a *regular part* of a.

Any regular part of a is a complement of $\imath(a)$ in $\Lambda[0,a]$, and hence it is a maximal regular element contained in a. Moreover, if m and n are regular parts of the same irregular element a then the sublattices $L[0,m]$ and $L[0,n]$ are isomorphic, since

$$L[0,m] = \Lambda[0,m] \cong \Lambda[\imath(a),a] \cong \Lambda[0,n] = L[0,n].$$

We remark that $L[0,m]$ is a complemented modular lattice.

Lemma (25.6). *Let $L \equiv \Lambda - S$ be a Wilcox lattice satisfying the condition (MC), and assume that there exist two elements $a, b \in L$ such that $0 < a < b < 1$.*

(25.6.1) *If an element a of L is neither zero nor regular then the following three statements are equivalent.*

(α) *a is an irregular element with a regular part m and $a \neq 1$.*

(β) *There exists an element $b \in L$ such that $a \underset{(m,n)}{\parallel} b$ with $m, n \in R$.*

(γ) *There exists an element $b \in L$ such that $a \underset{(m)}{<\!|} b$ with $m \in R$.*

(25.6.2) *$1 \in L$ is an irregular element with a regular part m if and only if there exist two irregular elements a_1 and a_2 such that $a_i \neq 1$ $(i = 1, 2)$, $a_1 \vee a_2 = 1$ and one of the following two statements holds:*

(1) $a_1 \underset{(m_1, m_2)}{\parallel} a_2$ *with* $m_1 \vee m_2 = m \in R$.

(2) a_1 *and* a_2 *have a common regular part* m.

Proof. (I) First we shall prove (25.6.1). (α)$\Rightarrow$(β). Let $a = m \sqcup \imath(a)$, $m \in R$ and $a \neq 1$. By the condition (MC) there exists a non-zero modular element n such that $a \wedge n = 0$. Since L is semicomplemented by (MC), we have $n \in R$ by (24.4). Hence it follows from (24.13) that there exists b such that $a \underset{(m,n)}{\parallel} b$.

The implication $(\beta)\Rightarrow(\gamma)$ is trivial and the implication $(\gamma)\Rightarrow(\alpha)$ follows from (24.10).

(II) Assume that $1=m\sqcup\imath(1)$ with $m\in R$. We remark that the relation $0\lessdot\imath(1)\lessdot 1$ in Λ contradicts the assumption of the lemma.

When $0\lessdot\imath(1)$, there exists $\lambda_1\in\Lambda$ such that $\imath(1)<\lambda_1<1$. Since $\Lambda[\imath(1),1]$ is isomorphic to $\Lambda[0,m]$ by the mapping $\lambda\to\lambda\sqcap m$, we have $0<\lambda_1\sqcap m<m$. Put $m_1=\lambda_1\sqcap m$ and let m_2 be a complement of m_1 in $\Lambda[0,m]$. Then $m_1\vee m_2=m_1\sqcup m_2=m\in R$ and hence $m_1,m_2\in R$. Putting $a_i=m_i\sqcup\imath(1)$ $(i=1,2)$, a_1 and a_2 are irregular elements such that $a_i\neq 1$ and $a_1\vee a_2=a_1\sqcup a_2=m_1\sqcup m_2\sqcup\imath(1)=m\sqcup\imath(1)=1$. Moreover, since $\Lambda[0,m]$ is isomorphic to $\Lambda[\imath(1),1]$ by the mapping $\lambda\to\lambda\sqcup\imath(1)$, we have

$$a_1\sqcap a_2=(m_1\sqcap m_2)\sqcup\imath(1)=\imath(1)\in S.$$

Hence $a_1 \underset{(m_1,m_2)}{\parallel} a_2$ by (24.10), whence (1) holds.

When there exists $u_1\in\Lambda$ such that $0<u_1<\imath(1)$, we take a complement u_2 of u_1 in $\Lambda[0,\imath(1)]$. Evidently $u_1,u_2\in S$. Putting $a_i=m\sqcup u_i$ $(i=1,2)$, a_1 and a_2 are irregular elements such that $a_i\neq 1$ and $a_1\vee a_2=a_1\sqcup a_2=m\sqcup\imath(1)=1$. Moreover (2) holds.

(III) Let a_1 and a_2 be irregular elements with $a_i\neq 1$ and $a_1\vee a_2=1$. If (1) holds, then by (24.10) we have

$$a_1\sqcap a_2\in S \quad\text{and}\quad a_i=m_i\sqcup(a_1\sqcap a_2) \qquad (i=1,2).$$

Hence we have

$$1=a_1\vee a_2=a_1\sqcup a_2=m_1\sqcup m_2\sqcup(a_1\sqcap a_2)=m\sqcup(a_1\sqcap a_2).$$

Since $m\in R$ and $a_1\sqcap a_2\in S$, 1 is an irregular element with a regular part m. If (2) holds, then $a_i=m\sqcup\imath(a_i)$ $(i=1,2)$, whence $1=a_1\sqcup a_2=m\sqcup\imath(a_1)\sqcup\imath(a_2)$. Since $\imath(a_1)\sqcup\imath(a_2)\in S$, 1 is an irregular element with a regular part m. □

Remark (25.7). Let $L\equiv\Lambda-S$ be a Wilcox lattice satisfying (MC), and assume that there exist $a,b\in L$ such that $0<a<b<1$. It follows from (25.2) and (25.6) that irregular elements of L can be defined without using imaginary elements.

Lemma (25.8). *Let $L\equiv\Lambda-S$ be a Wilcox lattice satisfying the condition (MC), and let a and b be irregular elements of L.*

When $a\neq 1$ and $b\neq 1$, the equality $\imath(a)=\imath(b)$ holds if and only if

(25.8.1) *there exist irregular elements a_i and regular elements m_i $(i=0,1,\dots,r)$ such that $a_0=a$, $a_r=b$ and $a_{i-1} \underset{(m_{i-1},m_i)}{\parallel} a_i$ for every $i=1,\dots,r$.*

When 1 *is irregular and* $a \neq 1$, *the equality* $\iota(a)=\iota(1)$ *holds if and only if*

(25.8.2) *there exists an irregular element* a' *such that* $a \vee a' = 1$ *and* $a \underset{(m,m')}{\parallel} a'$ *with* $m \vee m' \in R$.

Proof. (I) Let $a \neq 1$ and $b \neq 1$. If (25.8.1) holds, then by (24.11) we have $\iota(a_{i-1})=\iota(a_i)$ for every i, whence $\iota(a)=\iota(a_0)=\iota(a_r)=\iota(b)$.

Conversely, assume $\iota(a)=\iota(b)$. When $a \wedge b=0$, taking regular parts m and n of a and b respectively, we have $a \underset{(m,n)}{\parallel} b$ by (24.12).

When $a \wedge b \neq 0$, it follows from (MC) that there exist non-zero modular elements m_1, m_2 and m_3 such that

$$m_1 \wedge a=0, \quad m_2 \leqq a \wedge b \quad \text{and } m_3 \wedge b=0.$$

Then $m_i \in R$ for $i=1,2,3$. Let m_0 and m_4 be regular parts of a and b respectively. We put $a_i=m_i \sqcup \iota(a)$ for every $i=0,\dots,4$. Then $a_0=a$, $a_4=b$ and $a_2 \leqq a \wedge b$. Moreover it follows from (24.12) that

$$a_{i-1} \underset{(m_{i-1},m_i)}{\parallel} a_i \quad \text{for } i=1,2,3,4,$$

since $a_0 \wedge m_1=0$, $a_2 \wedge m_1 \leqq a \wedge m_1=0$, $a_2 \wedge m_3 \leqq b \wedge m_3=0$ and $a_4 \wedge m_3=0$. Therefore, in either case, (25.8.1) holds.

(II) Let 1 be irregular and $a \neq 1$. If (25.8.2) holds, then by (24.10) we have

$$a \sqcap a' \in S, \quad a=m \sqcup (a \sqcap a') \quad \text{and } a'=m' \sqcup (a \sqcap a').$$

Hence $1=a \sqcup a'=m \sqcup m' \sqcup (a \sqcap a')$. Since $m \sqcup m'=m \vee m' \in R$, we have $\iota(1)=a \sqcap a'=\iota(a)$.

Conversely, assume $\iota(a)=\iota(1)$. Let m be a regular part of a and let m' be a complement of a in Λ. Then we have $m \vee m' \in R$, since $\iota(1)$ is the greatest element of S and since

$$\begin{aligned}(m \vee m') \sqcap \iota(1) &= (m \sqcup m') \sqcap \iota(a) = (m \sqcup m') \sqcap a \sqcap \iota(a) \\ &= \{m \sqcup (m' \sqcap a)\} \sqcap \iota(a) = m \sqcap \iota(a) = 0.\end{aligned}$$

We put $a'=m' \sqcup \iota(a)$. Then a' is irregular and $a \vee a'=m \sqcup m' \sqcup \iota(a)$ $=a \sqcup m'=1$. Moreover $a \underset{(m,m')}{\parallel} a'$ since $a \wedge m' \leqq a \sqcap m'=0$. □

Theorem (25.9). *Let L be a Wilcox lattice satisfying the condition (MC), and assume that there exist elements a and b of L such that* $0<a<b<1$. *The modular extension* Λ *of L is uniquely determined up to isomorphism.*

Proof. (I) Let $L \equiv \Lambda_1 - S_1$ and $L \equiv \Lambda_2 - S_2$. By (25.1) we may assume that S_1 and S_2 are not empty. We shall show that there exists an isomorphism of Λ_1 onto Λ_2 which is the identity on L. It follows from (25.6) and (25.7) that if a is an irregular element of L and m is a regular part of a, then there are two imaginary parts $\iota_1(a) \in S_1$ and $\iota_2(a) \in S_2$, and then

(1) $$m \sqcup_1 \iota_1(a) = a = m \sqcup_2 \iota_2(a),$$

where $\sqcup_i$ denotes the join in Λ_i $(i=1,2)$.

(II) We take a modular element m of L with $0<m<1$. Since m is a regular element by (25.2), for $u \in S_1$ and $v \in S_2$ we can define

$$\varphi(u) = \iota_2(m \sqcup_1 u) \quad \text{and} \quad \psi(v) = \iota_1(m \sqcup_2 u).$$

If we put $a = m \sqcup_1 u$, then by (1) we have $a = m \sqcup_2 \iota_2(a) = m \sqcup_2 \varphi(u)$. Hence it holds that

(2) $$m \sqcup_1 u = m \sqcup_2 \varphi(u) \quad \text{for every } u \in S.$$

(Similarly, $m \sqcup_2 v = m \sqcup_1 \psi(v)$ for every $v \in S_2$.)

(III) We shall show that the mapping $\varphi: S_1 \to S_2$ is determined independent from the choice of m. For two modular elements m and n of L with $0<m, n<1$, we put $a = m \sqcup_1 u$ and $b = n \sqcup_1 u$ where $u \in S_1$. We need to show that $\iota_2(a) = \iota_2(b)$ holds even if $a \neq b$. When $a \neq 1$ and $b \neq 1$, the equalities $\iota_1(a) = u = \iota_1(b)$ imply the statement (25.8.1), and this implies $\iota_2(a) = \iota_2(b)$. When $a=1$ or $b=1$, $\iota_1(a) = \iota_1(b)$ implies $\iota_2(a) = \iota_2(b)$ by using (25.8.2). (The mapping ψ is also.)

(IV) For $u \in S_1$, it follows from (2) that

$$\psi\varphi(u) = \iota_1(m \sqcup_2 \varphi(u)) = \iota_1(m \sqcup_1 u) = u.$$

Similarly we have $\varphi\psi(v) = v$ for $v \in S_2$. Hence φ and ψ are mutually inverse one-to-one mappings between S_1 and S_2, and evidently they are order-preserving.

(V) We define a mapping $\bar{\varphi}$ of Λ_1 into Λ_2 as follows:

$$\bar{\varphi}(\lambda) = \begin{cases} \varphi(\lambda) & \text{when } \lambda \in S_1, \\ \lambda & \text{when } \lambda \in L. \end{cases}$$

Then $\bar{\varphi}$ is one-to-one and onto by (IV). Using (2), we can prove that $\bar{\varphi}$ is order-preserving in the same way as in the proof of (21.9). Therefore $\bar{\varphi}$ is an isomorphism of Λ_1 onto Λ_2 which is the identity on L. □

EXERCISE 25.1. Prove that if a Wilcox lattice L satisfies (MC) then L is left complemented.

EXERCISE 25.2. Let L be an upper continuous lattice satisfying (MC). Prove that the following statements are equivalent:

(α) L is $\perp$-symmetric.

(β) L is M-symmetric.

(γ) L is left complemented.

26. Modular Contractions and Modular Centers of Wilcox Lattices

Lemma (26.1). *Let m be a regular element of a Wilcox lattice $L \equiv \Lambda - S$. The set $I_m = \{m \sqcup u; u \in S\} \cup \{m\}$ is a relatively complemented modular sublattice of L, which is isomorphic to $S \cup \{0\}$.*

Proof. The mapping "$u \rightarrow m \sqcup u$" of $S \cup \{0\}$ onto I_m is one-to-one since $u = \iota(m \sqcup u)$. Let $u_1, u_2 \in S \cup \{0\}$. We have

$$(1) \qquad (m \sqcup u_1) \vee (m \sqcup u_2) = (m \sqcup u_1) \sqcup (m \sqcup u_2) = m \sqcup (u_1 \sqcup u_2).$$

Moreover $(m \sqcup u_1) \wedge (m \sqcup u_2) = (m \sqcup u_1) \sqcap (m \sqcup u_2)$, since $(m \sqcup u_1) \sqcap (m \sqcup u_2) \geqq m \in L$. Since $m \sqcap (u_1 \sqcup u_2) = 0$, by the isomorphism of $\Lambda[0, u_1 \sqcup u_2]$ and $\Lambda[m, m \sqcup u_1 \sqcup u_2]$, we have

$$(2) \qquad (m \sqcup u_1) \wedge (m \sqcup u_2) = (m \sqcup u_1) \sqcap (m \sqcup u_2) = m \sqcup (u_1 \sqcap u_2).$$

By (1) and (2), I_m is a sublattice of L which is isomorphic to $S \cup \{0\}$. Since $S \cup \{0\}$ is an ideal of Λ, I_m is relatively complemented and modular. $\square$

Definition (26.2). Let m be a regular element of a Wilcox lattice $L \equiv \Lambda - S$. If an element a of L is either regular or irregular then we put

$$\varphi_m(a) = m \sqcup \iota(a).$$

By the mapping $a \rightarrow \varphi_m(a)$, all regular elements and all irregular elements of L are transposed into the modular sublattice $I_m = \{m \sqcup u; u \in S\} \cup \{m\}$, preserving the order. We call I_m a *modular contraction* of L at m.

Remark (26.3). Let $L \equiv \Lambda - S$ be a Wilcox lattice. Evidently the following three statements are equivalent.

(α) $\varphi_m(a) = \varphi_m(b)$ for every $m \in R$.
(β) $\varphi_m(a) = \varphi_m(b)$ for some $m \in R$.
(γ) $\iota(a) = \iota(b)$.

If L satisfies the condition (MC), then it follows from (25.8) that $\varphi_m(a) = \varphi_m(b)$ is equivalent to (25.8.1) or (25.8.2). Compare with (19.6.1).

Theorem (26.4). *Let $L \equiv \Lambda - S$ be a Wilcox lattice where S is not empty. The set*

$$M(L) = \{a \in L; u < a \text{ for every } u \in S\} \cup \{0\}$$

is a modular sublattice of L.

Proof. (I) Let $a, b \in M(L)$. It is evident that $a \vee b \in M(L)$. To prove $a \wedge b \in M(L)$, we may assume $a \neq 0$ and $b \neq 0$. When $a \sqcap b \in S$, we have

$a \wedge b = 0 \in M(L)$. When $a \sqcap b \in L$, since $a \wedge b = a \sqcap b$, we have $u < a \wedge b$ for every $u \in S$, whence $a \wedge b \in M(L)$. Therefore $M(L)$ is a sublattice of L.

(II) We shall prove $(a,b)M$ in $M(L)$. When $a \sqcap b \in L$, since $(a,b)M$ in L and since $M(L)$ is a sublattice of L, we have $(a,b)M$ in $M(L)$. When $a \sqcap b \in S$, then $a \wedge b = 0$ and we have $a \sqcap b \leqq c$ for any $c \in M(L)$ with $0 < c \leqq b$. Hence by the modularity of Λ

$$c = c \sqcup (a \sqcap b) = (c \sqcup a) \sqcap b = (c \vee a) \sqcap b \geqq (c \vee a) \wedge b \geqq c.$$

Therefore $(c \vee a) \wedge b = c = c \vee (a \wedge b)$. Thus $(a,b)M$ in $M(L)$. ▯

Definition (26.5). Let $L \equiv \Lambda - S$ be a Wilcox lattice where S is not empty. We call the modular sublattice $M(L)$ in (26.4) the *modular center* of L. When S is empty, we may put $M(L) = L$.

Remark (26.6). If a Wilcox lattice $L \equiv \Lambda - S$ has the imaginary unit i, then it is easy to prove the following three statements.

(26.6.1) $$I_m = L[m, m \sqcup i] \cong \Lambda[0, i].$$

(26.6.2) The domain of φ_m is equal to $L - \{0\}$ (see (24.7.3)).

(26.6.3) $M(L) \cong \Lambda[i, 1]$ and $M(L)$ is complemented.

We remark that if L is an affine matroid lattice then the modular center $M(L)$ defined by (26.5) coincides with that defined by (22.4) (see (22.5)).

Lemma (26.7). *Let $L \equiv \Lambda - S$ be a Wilcox lattice where S is not empty.*

(26.7.1) *The following three statements are equivalent.*

(α) *There exists the imaginary unit for L.*

(β) *There exists an irregular element which belongs to $M(L)$.*

(γ) *Every non-zero element of $M(L)$ is irregular.*

(26.7.2) *Assume that there exists the imaginary unit for L. If a and b are two non-zero elements of $M(L)$ such that $a \wedge b = 0$ then $a \underset{(m,n)}{\parallel} b$ with $m, n \in R$.*

Proof. (I) We shall prove (26.7.1). (α)$\Rightarrow$(γ). Let i be the imaginary unit for L. If $a \in M(L)$ and $a \neq 0$, then since $i < a$, we can take a complement m of i in $\Lambda[0, a]$. Then m is regular since $m \sqcap i = 0$. Hence $a = m \sqcup i$ is irregular.

The implication (γ)$\Rightarrow$(β) is trivial.

(β)$\Rightarrow$(α). If there exists an irregular element a with $a \in M(L)$, then since $a \neq 0$, a contains all elements of S. Hence $\iota(a)$ is the greatest element of S by (24.8).

(II) Let $a, b \in M(L)$, $a \neq 0$, $b \neq 0$ and $a \wedge b = 0$. By (26.7.1), a and b are irregular. Hence

$$a = m \sqcup \iota(a) \quad \text{and} \quad b = n \sqcup \iota(b) \quad \text{with} \quad m, n \in R.$$

Since $a, b \in M(L)$, we have $\iota(a) = \iota(b) = i$. Therefore $a \underset{(m,n)}{\parallel} b$ by (24.12). □

Problem 5. Find a point-free parallel axiom such that a weakly modular symmetric lattice L satisfying (MC) satisfies this axiom if and only if L is a Wilcox lattice (cf. (20.10)). (One solution is given in S. Maeda [8].)

Problem 6. Find a lattice which may be placed in the part (?) of the following table.

	atomistic	general
complemented modular lattice	the lattice of subspaces of a vector space	the lattice of principal right ideals of a regular ring
weakly modular M-symmetric lattice (Wilcox lattice)	the lattice of affine subsets of a vector space	(?)
orthomodular M-symmetric lattice	the lattice of closed subspace of a Hilbert space	the lattice of projections of a von Neumann algebra

References for Chapter V

For Section 23 and 24: F. Maeda [6] and S. Maeda [8].

For Section 25: S. Maeda [8].

For Section 26: F. Maeda [6].

Chapter VI

Atomistic Symmetric Lattices with Duality

27. Modularity in DAC-lattices

Definition (27.1). A lattice L with 0 and 1 is called a *DAC-lattice* when both L and its dual L^* are AC-lattices, that is, atomistic lattices with the covering property. If L is a DAC-lattice then so is L^* evidently.

Remark (27.2). A matroid lattice may be defined as an upper continuous AC-lattice, and it is dual-atomistic by (13.1). Hence, the difference between a matroid lattice and a DAC-lattice is that the former is upper continuous and the latter has the dual covering property.

It follows from (14.1) that an upper continuous DAC-lattice is a modular matroid lattice. Conversely, a modular matroid lattice is an upper continuous DAC-lattice, since the modularity implies the dual-covering property.

Remark (27.3). When $a<b$ in a DAC-lattice L, the sublattice $L[a,b]$ is a DAC-lattice, because, by (8.18), $L[a,b]$ is an AC-lattice and the dual of $L[a,b]$ which is equal to $L^*[b,a]$ is also an AC-lattice.

Lemma (27.4). *An atomistic, dual-atomistic lattice L is a DAC-lattice if and only if*

(27.4.1) $\qquad a \wedge b \lessdot a$ *is equivalent to* $b \lessdot a \vee b$ *in* L.

Proof. Since L and L^* are atomistic, it follows from (7.10) that (27.4.1) holds if and only if both L and L^* have the covering property. ▯

Lemma (27.5). *If an AC-lattice L is dual-atomistic and finite-modular then L is a DAC-lattice.*

Proof. By (9.5) L^* is M-symmetric, whence L^* has the covering property by (7.7). Hence L is a DAC-lattice. ▯

Theorem (27.6). *Any DAC-lattice is M-symmetric and M*-symmetric. Moreover it is finite-modular.*

Proof. Let L be a DAC-lattice. By (27.4), both L and L^* satisfies (δ) of (9.5). Hence they are M*-symmetric and finite-modular. ▯

Remark (27.7). Let a and b be two elements of a DAC-lattice L and let $0<a<1$ and $0<b<1$. The following results follow from (9.3).

$(a,b)M$ in L if and only if the following statement holds:

(27.7.1) If h is a dual-atom of L and $h \geqq a \wedge b$ then there exist dual-atoms h_1 and h_2 of L such that $h \geqq h_1 \wedge h_2$, $h_1 \geqq a$ and $h_2 \geqq b$.

$(a,b)M^*$ in L if and only if the following statement holds:

(27.7.2) If p is an atom of L and $p \leqq a \vee b$ then there exist atoms q and r of L such that $p \leqq q \vee r$, $q \leqq a$ and $r \leqq b$.

Definition (27.8). An element a of a lattice L with 1 is called *dual-finite* when a is finite in the lattice L^*. In a lattice L with 0 and 1, the set of all elements which are either finite or dual-finite is denoted by $\check{\mathscr{F}}(L)$. In notation,

$$\check{\mathscr{F}}(L) = \mathscr{F}(L) \cup \mathscr{F}(L^*).$$

Lemma (27.9). *In a DAC-lattice L, if* $a \in \check{\mathscr{F}}(L)$ *then* $(a,b)M$, $(b,a)M$, $(a,b)M^*$ *and* $(b,a)M^*$ *for every* $b \in L$.

Proof. Since L and L^* are finite-modular, this lemma follows from (9.4). □

Theorem (27.10). *If L is a DAC-lattice, then* $\check{\mathscr{F}}(L)$ *is a sublattice of L which is a complemented modular DAC-lattice.*

Proof. (I) Let $a, b \in \check{\mathscr{F}}(L)$. We shall show that $a \wedge b \in \check{\mathscr{F}}(L)$. Since $\mathscr{F}(L)$ is an ideal of L by (8.8), we have $a \wedge b \in \mathscr{F}(L)$ when a or b is finite. When a and b are dual-finite, evidently $a \wedge b$ is dual-finite. Hence in either case, $a \wedge b \in \check{\mathscr{F}}(L)$. Similarly, $a \vee b \in \check{\mathscr{F}}(L)$.

(II) $\check{\mathscr{F}}(L)$ is modular by (27.9). $\check{\mathscr{F}}(L)$ is atomistic and dual-atomistic, since it contains all atoms and dual-atoms of L. Hence $\check{\mathscr{F}}(L)$ is a modular DAC-lattice.

(III) To prove that $\check{\mathscr{F}}(L)$ is complemented, by the duality, it suffices to show that every dual-finite element has a complement in $\check{\mathscr{F}}(L)$. When a is dual-finite, by the dual covering property there exist elements a_i $(i=0,1,\dots,n)$ such that

$$a = a_0 \lessdot a_1 \lessdot \cdots \lessdot a_n = 1.$$

Taking atoms p_i $(i=1,\dots,n)$ such that $p_i \leqq a_i$ and $p_i \not\leqq a_{i-1}$, we have

(1) $$(a \vee p_1 \vee \cdots \vee p_{i-1}) \wedge p_i = a_{i-1} \wedge p_i = 0 \quad \text{for } i=1,\dots,n,$$

and $a \vee p_1 \vee \cdots \vee p_n = 1$. In $\check{\mathscr{J}}(L)$, $x \wedge y = 0$ is equivalent to $x \perp y$, since $\check{\mathscr{J}}(L)$ is modular (see (3.1)). Hence, by (2.5), (1) implies $(a, p_1, \ldots, p_n) \perp$ in $\check{\mathscr{J}}(L)$, whence

$$a \wedge (p_1 \vee \cdots \vee p_n) = 0.$$

Therefore, the finite element $p_1 \vee \cdots \vee p_n$ is a complement of a. □

Lemma (27.11). *Let a and b be elements of an AC-lattice L. If $(a,x)M$ for every $x \succ b$ then $(a,b)M^*$.*

Proof. Let $c \geqq b$ and let p be an atom such that $p \leqq c \wedge (a \vee b)$. We shall prove that $p \leqq (c \wedge a) \vee b$. This is evident when $p \leqq b$. When $p \nleqq b$, putting $x = b \vee p$, we have $b \prec x \leqq c$. Since $(a,x)M$,

$$p \leqq x \wedge (a \vee b) = (b \vee a) \wedge x = b \vee (a \wedge x) \leqq (c \wedge a) \vee b.$$

Hence, we have $c \wedge (a \vee b) \leqq (c \wedge a) \vee b$, since L is atomistic. Thus $(a,b)M^*$ holds. □

Lemma (27.12). *In a finite-modular AC-lattice L, if $(a,b)M^*$ then $(a \vee x, b \vee y)M^*$ for all finite elements x and y.*

Proof. By the dual property of (1.5.1),

(1) $\quad (a,b)M^*$ and $(a \vee b, c)M^*$ together imply $(a, b \vee c)M^*$.

Let $(a,b)M^*$. If y is a finite element, then since $(a \vee b, y)M^*$ by (9.4), we have $(a, b \vee y)M^*$ by (1). Similarly, since L is M^*-symmetric, $(a, b \vee y)M^*$ implies $(a \vee x, b \vee y)M^*$ for every finite element x. □

Lemma (27.13). *Let a and b be elements of a finite-modular AC-lattice L such that $(a,b)M$ and $(a,b)M^*$. If $a \wedge b \leqq x \leqq b$ then $(a,x)M^*$.*

Proof. We may assume $a \neq 0$ and $x \neq 0$. Let p be an atom such that $p \leqq a \vee x$. By (9.3) it suffices to show that there exist atoms q and r such that $p \leqq q \vee r$, $q \leqq a$ and $r \leqq x$. Since $p \leqq a \vee b$ and $(a,b)M^*$, there exist atoms q_1 and r_1 such that $p \leqq q_1 \vee r_1$, $q_1 \leqq a$ and $r_1 \leqq b$. When $p = q_1$, then $q = q_1$ and any atom $r \leqq x$ may be used.

When $p \neq q_1$, by the exchange property (7.8.1) we have $r_1 \leqq p \vee q_1 \leqq a \vee x$. Since $(a,b)M$ and $a \wedge b \leqq x \leqq b$,

$$r_1 \leqq (x \vee a) \wedge b = x \vee (a \wedge b) = x.$$

Hence $q = q_1$ and $r = r_1$ have the desired property. □

Theorem (27.14). *Let a and b be elements of a DAC-lattice L. The following three statements are equivalent.*

(α) *$(a,b)M$ and $(a,b)M^*$.*
(β) *$(a,x)M$ for every $x \succ b$.*
(γ) *$(a,x)M^*$ for every $x \prec b$.*

Proof. (I) We shall prove that (γ) implies (α). When $b=0$, evidently (α) holds. When $b\neq 0$, since $L[0,b]$ is a DAC-lattice by (27.3), there exists an element c such that $c \lessdot b$. Then there exists an atom p such that $b=c\vee p$. Since $(a,c)M^*$ by (γ), we have $(a,b)M^*$ by (27.12).

Moreover, by (γ), in L^* we have $(a,x)M$ for every $x \gtrdot b$. Hence, by (27.11), $(a,b)M^*$ in L^*, whence $(a,b)M$ in L. Therefore (γ) implies (α).

If (β) holds, then (γ) holds in L^* and hence (α) holds in L^*. Therefore (α) holds in L.

(II) We shall prove that (α) implies (β). Let $x \gtrdot b$. When $x\leqq a\vee b$, then in L^* we have $a\wedge b\leqq x\leqq b$. Hence, by (27.13), (α) implies $(a,x)M^*$ in L^*, whence $(a,x)M$ in L. When $x\nleqq a\vee b$, we take an atom p such that $x=b\vee p$. Then $p\wedge(a\vee b)=0$, since otherwise we would have $x\leqq a\vee b$. Since L is M-symmetric, we have $(b,a)M$ and $(p,b\vee a)M$. Moreover, $p\wedge(b\vee a)=0\leqq b$. Hence by (1.6) we have $(p\vee b,a)M$. Therefore (α) implies (β).

By the duality, (α) implies (γ). ▯

Corollary (27.15). *Let L be a DAC-lattice.*

(27.15.1) *In L the following three statements are equivalent.*

(α) $(a,b)M$ *implies* $(a,b)M^*$ (L is cross-symmetric).
(β) *If* $(a,b)M$ *and* $b\lessdot c$, *then* $(a,c)M$.
(γ) *If* $(a,b)M$ *then* $(a\vee x, b\vee y)M$ *for all finite elements x and y.*

(27.15.2) *In L the following three statements are equivalent.*

(α^*) $(a,b)M^*$ *implies* $(a,b)M$ (L is dual cross-symmetric).
(β^*) *If* $(a,b)M^*$ *and* $b\gtrdot c$ *then* $(a,c)M^*$.
(γ^*) *If* $(a,b)M^*$ *then* $(a\wedge x, b\wedge y)M^*$ *for all dual-finite elements x and y.*

Proof. (I) If $(a,b)M$ holds, then, by the equivalence of (α) and (β) in (27.14), $(a,b)M^*$ holds if and only if $(a,x)M$ holds for every $x\gtrdot b$. Hence (α) and (β) are equivalent.

(II) (γ)$\Rightarrow$(β). If $b\lessdot c$ then there exists an atom p such that $c=b\vee p$. Hence (γ) implies (β).

(β)$\Rightarrow$(γ). Assume $(a,b)M$. If y is a finite element, then by the covering property there exist elements $b_1,\ldots,b_n$ such that

$$b\lessdot b_1\lessdot b_2\lessdot \cdots \lessdot b_n=b\vee y.$$

Applying (β) successively, we have $(a,b\vee y)M$. Since L is M-symmetric, similarly $(a,b\vee y)M$ implies $(a\vee x, b\vee y)M$ for any finite element x. Hence (β) implies (γ).

(III) The equivalence of (α^*), (β^*) and (γ^*) can be proved in a similar way. ▯

Remark (27.16). Let L be a DAC-lattice. Since L is a finite-modular AC-lattice, it follows from (15.5), (15.10) and (15.11) that the set $L(\Omega(L))$ of all subspaces of the atom space $\Omega(L)$ forms a modular matroid lattice, ordered by set-inclusion, and that the mapping $a \to \omega(a)$ of L into $L(\Omega(L))$ has the following properties.

(27.16.1) ω is one-to-one and order-preserving.

(27.16.2) $\omega(0)=0$ and $\omega(1)=1$.

(27.16.3) $\omega(\bigwedge_\alpha a_\alpha) = \bigwedge_\alpha \omega(a_\alpha)$ (if $\bigwedge_\alpha a_\alpha$ exists).

(27.16.4) $\omega(a \vee b) \geqq \omega(a) \vee \omega(b)$ for every a and b, and equality holds if and only if $(a,b)M^*$ holds in L.

(27.16.5) $a \to \omega(a)$ is an isomorphism of $\mathscr{J}(L)$ onto $\mathscr{J}(L(\Omega(L)))$.

(27.16.6) If $a \lessdot b$ in L then $\omega(a) \lessdot \omega(b)$ in $L(\Omega(L))$.

Lemma (27.17). *Let L be a DAC-lattice.*

(27.17.1) *If h is a dual-atom of L then $\omega(h)$ is a dual-atom of $L(\Omega(L))$.*

(27.17.2) *If a is a dual-finite element of L then $\omega(a)$ is a dual-finite element of $L(\Omega(L))$.*

(27.17.3) *The image $\omega(\mathscr{J}(L^*))$ is a dual-ideal of $L(\Omega(L))$.*

Proof. (I) If $h \lessdot 1$ in L then by (27.16.6) $\omega(h) \lessdot \omega(1)=1$ in $L(\Omega(L))$. Hence $\omega(h)$ is a dual-atom of $L(\Omega(L))$.

(II) If a is a dual-finite element then there exist dual-atoms $h_1,\dots,h_n$ such that $a = h_1 \wedge \cdots \wedge h_n$. By (27.16.3) we have $\omega(a) = \omega(h_1) \wedge \cdots \wedge \omega(h_n)$. Since $\omega(h_i)$ is a dual-atom for every i, $\omega(a)$ is dual-finite.

(III) Since L^* is an AC-lattice, $\mathscr{J}(L^*)$ is a dual-ideal of L. If $a, b \in \mathscr{J}(L^*)$, then since $a \wedge b \in \mathscr{J}(L^*)$, we have

$$\omega(a) \wedge \omega(b) = \omega(a \wedge b) \in \omega(\mathscr{J}(L^*)).$$

We shall show that if $a \in \mathscr{J}(L^*)$ and $\omega(a) < \omega$ in $L(\Omega(L))$ then there exists $b \in \mathscr{J}(L^*)$ such that $\omega = \omega(b)$. Since $\omega(a)$ is dual-finite and since $L(\Omega(L))$ has the dual covering property, there exist $\omega_1, \dots, \omega_n \in L(\Omega(L))$ such that

$$\omega(a) \lessdot \omega_1 \lessdot \omega_2 \lessdot \cdots \lessdot \omega_n = \omega.$$

We take an atom $p \in \omega_1$ such that $p \notin \omega(a)$, and we put $a_1 = a \vee p$. Since

$$\omega(a) < \omega(a_1) = \omega(a) \vee \omega(p) \leqq \omega_1,$$

we have $\omega_1 = \omega(a_1)$. Continuing this process, there exists an element $b \in L$ such that $\omega = \omega(b)$. Since $a < b$, we have $b \in \mathscr{J}(L^*)$. Therefore $\omega(\mathscr{J}(L^*))$ is a dual-ideal of $L(\Omega(L))$. □

Exercise 27.1. Show that any complemented modular atomic lattice is a DAC-lattice.

28. Complete DAC-lattices

Lemma (28.1). *Let L be a lattice. The completion by cuts of L and that of L^* are dual-isomorphic by the mutually inverse mappings $X \to X^u$ and $X \to X^l$.*

Proof. A subset X of L belongs to $\overline{L^*}$ (the completion by cuts of L^*) if and only if $X = X^{lu}$. Since $X^{ulu} = X^u$ and $X^{lul} = X^l$ for every X, the mappings $X \to X^u$ and $X \to X^l$ are mutually inverse mappings between $\overline{L}$ and $\overline{L^*}$. Hence $\overline{L}$ and $\overline{L^*}$ are dual-isomorphic. □

Theorem (28.2). *If L is a DAC-lattice then its completion by cuts $\overline{L}$ is also a DAC-lattice.*

Proof. By (12.7), both $\overline{L}$ and $\overline{L^*}$ are AC-lattices. Hence L is a DAC-lattice by (28.1). □

Theorem (28.3). *If L is a complete DAC-lattice then L is isomorphic to the completion by cuts of the sublattice $\check{\mathscr{I}}(L)$.*

Proof. (I) For every element a of L, we put

$$X_a = \{x \in \check{\mathscr{I}}(L); x \leqq a\}.$$

We shall show that $X_a \in \overline{\check{\mathscr{I}}(L)}$. In $\check{\mathscr{I}}(L)$, let $x \in X_a^{ul}$. If h is a dual-atom of L with $h \geqq a$, then $x \leqq h$ since $h \in X_a^u$ in $\check{\mathscr{I}}(L)$. Since L is dual-atomistic, we have $x \leqq a$, whence $x \in X_a$. Therefore $X_a \in \overline{\check{\mathscr{I}}(L)}$.

(II) The mapping $a \to X_a$ of L into $\overline{\check{\mathscr{I}}(L)}$ is evidently order-preserving. It is one-to-one, because if $a \nleqq b$, then there exists an atom p with $p \leqq a$ and $p \nleqq b$ and then we have $X_a \nleqq X_b$.

(III) We shall show that this mapping is onto. Let $X \in \overline{\check{\mathscr{I}}(L)}$ and put $a = \bigvee(x; x \in X)$. If $x \in X$, then $x \in X_a$ since $x \leqq a$. Conversely, if $x \in X_a$, then for any element y of X^u we have $x \leqq a \leqq y$. Hence $x \in X^{ul} = X$. Therefore $X = X_a$.

By (II) and (III), L is isomorphic to $\overline{\check{\mathscr{I}}(L)}$. □

Remark (28.4). The completion by cuts of a modular lattice is not necessarily modular. For instance, if a complete DAC-lattice L is not modular, then, though $\check{\mathscr{I}}(L)$ is a modular lattice by (27.10), its completion by cuts is not modular by (28.3).

Remark (28.5). It follows from (28.3) that two complete DAC-lattices L_1 and L_2 are isomorphic if $\check{\mathscr{I}}(L_1)$ and $\check{\mathscr{I}}(L_2)$ are isomorphic.

Remark (28.6). Let L be a DAC-lattice. Since L is finite-modular, in L the perspectivity of atoms is transitive by (11.7). Moreover, L is SSC* since it is dual-atomistic. Hence by (10.4.3) if two atoms of L are projective then they are perspective.

If L is moreover complete, then L is a Z-lattice by (10.7).

Theorem (28.7). *Let a and b be elements of a complete DAC-lattice L. The following four statements are equivalent.*

(α) $a \triangledown b$.
(β) $e(a) \wedge e(b) = 0$.
(γ) *There do not exist atoms p and q such that*

$$p \sim q, \quad p \leqq a \quad \text{and} \quad q \leqq b.$$

(δ) *There do not exist non-zero elements a_1 and b_1 such that*

$$a_1 \sim b_1, \quad a_1 \leqq a \quad \text{and} \quad b_1 \leqq b.$$

Proof. This theorem follows from (10.10), since L satisfies the assumptions of (10.10) by (28.6). ▯

Theorem (28.8). *Let L be a complete DAC-lattice.*

(28.8.1) *L is the direct sum of irreducible sublattices $\{L[0, z_\alpha]; \alpha \in I\}$ where $z_\alpha \in Z(L)$ for every $\alpha \in I$.*

(28.8.2) *Two atoms p and q of L are contained in the same summand $L[0, z_\alpha]$ if and only if they are perspective.*

(28.8.3) *L is irreducible if and only if any two atoms of L are perspective.*

Proof. This theorem follows from (10.14), since $p \approx q$ is equivalent to $p \sim q$ by (28.6). ▯

Remark (28.9). The results (28.7) and (28.8) on a complete DAC-lattice coincide with (13.5) and (13.6) on a matroid lattice.

EXERCISE 28.1. Let L be a dual-atomistic lattice. Prove that (1) $\bar{L}$ is dual-atomistic, (2) $X \in \bar{L}$ is a dual-atom of $\bar{L}$ if and only if there exists a dual-atom h of L such that $X = J_h$, and (3) $\mathscr{J}(\overline{L^*}) = \{J_a; a \in \mathscr{J}(L^*)\}$.

29. Orthocomplemented Lattices and Orthomodular Lattices

Definition (29.1). A lattice L with 0 and 1 is called *orthocomplemented* when there is a mapping $a \to a^\perp$ of L into itself satisfying the following three conditions:

(29.1.1) $a^\perp$ is a complement of a.

(29.1.2) $a \leqq b$ implies $a^\perp \geqq b^\perp$.

(29.1.3) $a^{\perp\perp} = a$ for every a.

We call $a^\perp$ the *orthocomplement* of a. When $a \leqq b^\perp$, we say that a and b are *orthogonal* and write $a \perp b$.

Remark (29.2). By (29.1.2) and (29.1.3), the orthocomplementation $a \to a^\perp$ is a dual-isomorphism of L onto itself. Hence L and L^* are isomorphic by the orthocomplementation. From this fact, we have

$$0^\perp = 1, \quad 1^\perp = 0, \quad (a \vee b)^\perp = a^\perp \wedge b^\perp \text{ and } (a \wedge b)^\perp = a^\perp \vee b^\perp.$$

We remark that the condition (29.1.1) may be replaced by

(29.2.1) $$a \wedge a^\perp = 0,$$

since this implies $a \vee a^\perp = (a^\perp \wedge a)^\perp = 0^\perp = 1$.

Lemma (29.3). *An orthocomplemented lattice L is a semi-orthocomplemented lattice when $a \perp b$ is defined as in* (29.1). *If L is moreover complete, then this relation "$\perp$" is ortho-continuous.*

Proof. (I) If $a \perp a$, then by definition we have $a \leqq a^\perp$, whence $a = a \wedge a^\perp = 0$. Hence the axiom ($\perp$1) in (2.1) holds. ($\perp$2) holds, since $a \leqq b^\perp$ implies $a^\perp \geqq b^{\perp\perp} = b$. ($\perp$3) holds, since $a \perp b$ and $a_1 \leqq a$ together imply $a_1 \leqq a \leqq b^\perp$.

If $a \perp b$ and $a \vee b \perp c$, then since $b \leqq a^\perp$ and $c \leqq (a \vee b)^\perp \leqq a^\perp$, we have $b \vee c \leqq a^\perp$, whence $a \perp b \vee c$. Thus ($\perp$4) holds.

Evidently $a^\perp$ is a semi-orthocomplement of a.

(II) Let L be complete and let $a_\delta \uparrow a$ and $a_\delta \perp b$ for every $\delta \in D$. Since $a_\delta \leqq b^\perp$ for every $\delta \in D$, we have $a = \bigvee(a_\delta; \delta \in D) \leqq b^\perp$, whence $a \perp b$. Therefore the relation "$\perp$" satisfies (2.15.1). ☐

Remark (29.4). If an orthocomplemented lattice is $\perp$-symmetric, then it has two kinds of semi-orthogonality given by (3.1) and (29.3). In spite of this, in any orthocomplemented lattice we shall use the symbol "$\perp$" only for orthogonality (except in the word "$\perp$-symmetric").

Remark (29.5). The orthogonality relation satisfies the following axiom which is stronger than ($\perp$4).

($\perp$4′) $$a \perp b, \quad a \perp c \quad \text{imply } a \perp b \vee c.$$

If a semi-orthogonality relation satisfies ($\perp$4′), then a family S of elements is a semi-orthogonal family whenever $x \perp y$ holds for all $x, y \in S$ $(x \neq y)$.

In an orthocomplemented lattice, a family S of elements is called an *orthogonal family* when any two elements of S are orthogonal.

Remark (29.6). Let L be an orthocomplemented lattice. Since $a \to a^\perp$ is a dual-automorphism of L, we have

(29.6.1) $(a, b)M^*$ is equivalent to $(a^\perp, b^\perp)M$.

(29.6.2) L is M-symmetric if and only if L is M*-symmetric.

(29.6.3) $(a, b, c)D^*$ is equivalent to $(a^\perp, b^\perp, c^\perp)D$.

Moreover, it is easy to show that

(29.6.4) $$a \triangledown b \quad \text{implies } a \perp b.$$

Definition (29.7). An orthocomplemented lattice L is called *O-symmetric* when in L

(29.7.1) $$(a,b)M \quad \text{implies } (b^\perp, a^\perp)M.$$

Theorem (29.8). *If an orthocomplemented lattice L is O-symmetric, then L is M-symmetric and M*-symmetric. Moreover, in L, the four relations $(a,b)M$, $(b,a)M$, $(a,b)M^*$ and $(b,a)M^*$ are equivalent.*

Proof. If L is O-symmetric, then L satisfies (1.9.1) by (29.6.1). Hence L is M-symmetric, and it is M*-symmetric by (29.6.2). Moreover, it is evident that $(a,b)M$ and $(b,a)M^*$ are equivalent by (29.7.1). □

Lemma (29.9). *An element z of an orthocomplemented lattice L is a central element if and only if*

(29.9.1) $$a=(a\wedge z)\vee(a\wedge z^\perp) \quad \textit{for every } a\in L.$$

Proof. If (29.9.1) holds, then, taking orthocomplements, we have

$$a=(a\vee z)\wedge(a\vee z^\perp) \quad \text{for every } a\in L.$$

Hence z is central by (4.13). If z is central then (29.9.1) holds evidently. □

Lemma (29.10). *If L is an orthocomplemented lattice and if a is a non-zero element of L such that $(a^\perp, a)M$, then the sublattice $L[0,a]$ is orthocomplemented.*

Proof. We shall show that the mapping $x\to x^\perp\wedge a$ is an orthocomplementation in $L[0,a]$. Evidently the conditions (29.2.1) and (29.1.2) are satisfied. Let $x\in L[0,a]$. Since $(a^\perp,a)M$ we have

$$(x^\perp\wedge a)^\perp\wedge a=(x\vee a^\perp)\wedge a=x\vee(a^\perp\wedge a)=x.$$

Hence (29.1.3) is satisfied. □

Lemma (29.11). *If L is an orthocomplemented lattice then its completion by cuts $\bar{L}$ is also orthocomplemented, and then $(J_a)^\perp=J_{a^\perp}$ for every $a\in L$.*

Proof. (I) For every $X\in\bar{L}$, we put

$$X^\perp=\{y\in L; y\perp x \text{ for every } x\in X\}.$$

We shall show that $X^\perp\in\bar{L}$. Let $a\in(X^\perp)^{ul}$. If $x\in X$ and $y\in X^\perp$, then $x\perp y$, whence $y\leqq x^\perp$. Hence $x^\perp\in(X^\perp)^u$ for every $x\in X$. Therefore, for every $x\in X$, we have $a\leqq x^\perp$, whence $a\perp x$. Thus $a\in X^\perp$.

(II) We shall show that $X \to X^{\perp}$ is an orthocomplementation in $\bar{L}$. If $x \in X \wedge X^{\perp}$, then since $x \perp x$, we have $x = x \wedge x^{\perp} = 0$. Hence $X \wedge X^{\perp} = \{0\}$. It is evident that $X \leqq Y$ implies $X^{\perp} \geqq Y^{\perp}$.

Evidently $X \subset X^{\perp\perp}$. Conversely, if $x \in X^{\perp\perp}$, then for every $y \in X^{u}$, since $y^{\perp} \in X^{\perp}$, we have $x \perp y^{\perp}$, whence $x \leqq y$. Hence $x \in X^{ul} = X$.

Therefore $X \to X^{\perp}$ is an orthocomplementation, and we have

$$(J_a)^{\perp} = \{y \in L;\ y \perp a\} = \{y \in L;\ y \leqq a^{\perp}\} = J_{a^{\perp}}. \quad \square$$

Definition (29.12). An orthocomplemented lattice L is called *orthomodular* when in L

(29.12.1) $$a \perp b \quad \text{implies } (a, b) M.$$

Theorem (29.13). *Let L be an orthocomplemented lattice. The following five statements are equivalent.*

(α) *L is orthomodular.*
(β) *$(a, a^{\perp}) M$ for every $a \in L$.*
(γ) *$(a, a^{\perp}) M^*$ for every $a \in L$.*
(δ) *If $a \leqq b$ then $b = a \vee (b \wedge a^{\perp})$.*
(ε) *If $a \leqq b$ then there exists $c \in L$ such that*

$$a \perp c \quad \text{and } a \vee c = b.$$

Proof. The implications $(\alpha) \Rightarrow (\beta)$ and $(\delta) \Rightarrow (\varepsilon)$ are evident. (β) and (γ) are equivalent by (29.6.1).

$(\gamma) \Rightarrow (\delta)$. If $a \leqq b$, then since $(a^{\perp}, a) M^*$ by (γ), we have

$$b = b \wedge (a^{\perp} \vee a) = (b \wedge a^{\perp}) \vee a.$$

Finally, the implication $(\varepsilon) \Rightarrow (\alpha)$ follows from (2.9). $\square$

Remark (29.14). By (ε) in (29.13), an orthomodular lattice may be called a *relatively orthocomplemented* lattice. (It is relatively complemented by (2.10).) We remark that a relative orthocomplement c of a in b is uniquely determined and it is $b \wedge a^{\perp}$, since

$$c = c \vee (a \wedge a^{\perp}) = (c \vee a) \wedge a^{\perp} = b \wedge a^{\perp}.$$

Lemma (29.15). *If $a < b$ in an orthomodular lattice L, then the sublattice $L[a, b]$ is also orthomodular.*

Proof. (I) We shall prove that the mapping $x \to x' = (x^{\perp} \vee a) \wedge b$ is an orthocomplementation in $L[a, b]$. It is evident that $x \leqq y$ implies $x' \geqq y'$. We need to show that $x \wedge x' = a$ and $x'' = x$ for every $x \in L[a, b]$. Since $(x^{\perp}, x) M$, we have

$$x \wedge x' = x \wedge (x^{\perp} \vee a) \wedge b = (a \vee x^{\perp}) \wedge x = a \vee (x^{\perp} \wedge x) = a.$$

Since $x = a \vee (x \wedge a^{\perp})$ by (δ) of (29.13) and since $(b^{\perp}, b)M$, we have

$$x'' = \{(x')^{\perp} \vee a\} \wedge b = \{(x \wedge a^{\perp}) \vee b^{\perp} \vee a\} \wedge b$$
$$= (x \vee b^{\perp}) \wedge b = x \vee (b^{\perp} \wedge b) = x.$$

Hence $L[a,b]$ is orthocomplemented.

(II) To prove that $L[a,b]$ is orthomodular, it suffices to show that if $a \leqq x \leqq y \leqq b$, then $y = x \vee (y \wedge x')$. Since

$$y \wedge x' = y \wedge (x^{\perp} \vee a) \wedge b = y \wedge (x^{\perp} \vee a) \geqq y \wedge x^{\perp},$$

we have

$$y = x \vee (y \wedge x^{\perp}) \leqq x \vee (y \wedge x') \leqq y. \quad \square$$

Lemma (29.16). *Any complete orthomodular lattice is a Z-lattice.*

Proof. Since an orthomodular lattice L is relatively complemented, it follows from (5.14) that if L is complete then it is a Z-lattice. $\square$

Theorem (29.17). *If L is an orthomodular lattice then the following three statements are equivalent.*

(α) *L is* $\perp$*-symmetric.*
(β) *L is M-symmetric.*
(γ) *L is M*-symmetric.*

Proof. (β) and (γ) are equivalent by (29.6.2). The implication (β)$\Rightarrow$(α) is trivial.

(α)$\Rightarrow$(β). By (β) and (γ) in (29.13), L satisfies the condition (1.14.1). Hence (α) implies (β). $\square$

EXERCISE 29.1. Find an orthocomplemented $\perp$-symmetric lattice which is not M-symmetric (cf. Exercise 1.2).

30. Orthocomplemented AC-lattices

Remark (30.1). If L is an orthocomplemented lattice, then L and L^* are isomorphic by the orthocomplementation. Hence, an orthocomplemented AC-lattice is a DAC-lattice.

By (10.7), a complete orthocomplemented atomistic lattice is a Z-lattice.

Theorem (30.2). *Let L be an orthocomplemented atomistic lattice. The following four statements are equivalent.* (Compare with (7.15) and (29.17).)

(α) *L has the covering property.*
(β) *L is* $\perp$*-symmetric.*
(γ) *L is M-symmetric.*
(δ) *L is M*-symmetric.*

Proof. (γ) implies (β) evidently, and (β) implies (α) by (7.7). If L has the covering property, then L is a DAC-lattice by (30.1). Hence L is M-symmetric by (27.6). (γ) and (δ) are equivalent by (29.6.2). □

Lemma (30.3). *Let L be an orthocomplemented AC-lattice. The set* $\check{\mathscr{J}}(L)$ *of all finite elements and all dual-finite elements forms a sublattice of L which is an orthocomplemented atomistic modular lattice.*

Proof. It follows from (27.10) that $\check{\mathscr{J}}(L)$ is a sublattice of L and is atomistic and modular. An element a of L is dual-finite if and only if $a^{\perp}$ is finite. Hence $\check{\mathscr{J}}(L)$ is orthocomplemented by the mapping $a \rightarrow a^{\perp}$. □

Remark (30.4). The completion by cuts of an orthocomplemented AC-lattice is also an orthocomplemented AC-lattice by (12.7) and (29.11). But the completion by cuts of an orthomodular AC-lattice is not necessarily orthomodular. For instance, if an orthocomplemented AC-lattice L is not orthomodular (cf. (34.2) and (34.9)), then $\check{\mathscr{J}}(L)$ is an orthocomplemented modular AC-lattice by (30.3), while its completion by cuts is not orthomodular, since it is isomorphic to L by (28.3).

Theorem (30.5). *An orthocomplemented AC-lattice L is O-symmetric if and only if in L*

(30.5.1) $(a,b)M$ *and* $b \lessdot c$ *together imply* $(a,c)M$.

Proof. This is a direct consequence of (27.15.1), since the statement (α) is equivalent to O-symmetry of L. □

Remark (30.6). The results (28.7) and (28.8) are valid for a complete orthocomplemented AC-lattice, and in (28.8.1) each summand $L[0, z_{\alpha}]$ is orthocomplemented by (29.10).

Lemma (30.7). *An orthocomplemented AC-lattice L is orthomodular if and only if L satisfies the following condition:*

(30.7.1) *If* $a \in L$ *with* $0 < a < 1$ *and if p is an atom of L then there exist two atoms q and r such that* $p \leqq q \vee r$, $q \leqq a$ *and* $r \leqq a^{\perp}$.

Proof. Since L is a DAC-lattice it follows from (27.7) that (30.7.1) is equivalent to the condition that $(a, a^{\perp})M^*$ holds for every $a \in L$. Hence (30.7.1) is equivalent to orthomodularity by (29.13). □

Remark (30.8). In a complete orthomodular atomic lattice L, every non-zero element a has an orthogonal base; precisely speaking, there exists an orthogonal family of atoms p_{α} ($\alpha \in I$) such that $a = \bigvee(p_{\alpha}; \alpha \in I)$. This can be proved using Zorn's lemma as in the proof of (8.16).

Theorem (30.9). (Comparability theorem). *Let a and b be elements of a complete orthomodular AC-lattice L. There exist elements a', a'', b' and b'' such that*

$$a=a' \vee a'', \quad a' \perp a'', \quad b=b' \vee b'', \quad b' \perp b'',$$
$$a' \overset{p}{\sim} b', \quad \text{and} \quad e(a'') \wedge e(b'')=0.$$

Moreover $e(a')=e(b')=e(a) \wedge e(b)$.

Proof. Since L is relatively orthocomplemented, this theorem can be proved just as we proved (13.7), using orthogonal families instead of semi-orthogonal families, and using (28.7) instead of (13.5). ▯

PROBLEM 7. Is there an orthocomplemented AC-lattice which is not O-symmetric? Is there an orthomodular AC-lattice which is not O-symmetric?

References for Chapter VI

For Section 27: S. Maeda [3] and [6], G. W. Mackey [1].

For Section 28: J. E. McLaughlin [1], S. Maeda [6].

For Section 29: E.A. Schreiner [1], M.D. MacLaren [1], F. Maeda [3].

For Section 30: S. Maeda [3], M.D. MacLaren [1], U. Sasaki [4].

Chapter VII

Atomistic Lattices of Subspaces of Vector Spaces

31. The Lattice of Closed Subspaces of a Locally Convex Space

Let E be a vector space over a division ring K. We proved in (16.9) that the set $L(E)$ of all subspaces of E forms an irreducible modular matroid lattice. In this lattice the meet of subspaces is their set-intersection and the join of $A, B \in L(E)$ is the linear sum $A+B = \{x+y : x \in A, y \in B\}$.

Lemma (31.1). *Let E be a vector space over a division ring K. If a family L of subspaces of E satisfies the following three conditions:*

(31.1.1) $\{0\} \in L$ *and* $E \in L$,

(31.1.2) *if* $A_\alpha \in L$ *for every* $\alpha \in I$, *then* $\bigcap(A_\alpha; \alpha \in I) \in L$,

and

(31.1.3) *if* $A \in L$ *then* $A + Kx \in L$ *for every* $x \in E$,

then

(31.1.4) *L forms an irreducible complete finite-modular AC-lattice, ordered by set-inclusion.*

(31.1.5) *For* $A, B \in L$, $(A, B) M^*$ *in L if and only if* $A + B \in L$, *or equivalently,* $A \vee B = A + B$.

(31.1.6) *The lattice* $L(\Omega(L))$ *is isomorphic to* $L(E)$.

Proof. Put $\Lambda = L(E)$. Since the meet $\sqcap_\alpha A_\alpha$ in Λ is equal to $\bigcap_\alpha A_\alpha$ and since the join $A \sqcup B$ in Λ is equal to $A+B$, by the above three conditions the subset L of Λ satisfies the three conditions in (15.15). Hence L forms a complete AC-lattice and $L(\Omega(L))$ is isomorphic to $\Lambda = L(E)$. Since Λ is modular, it follows from (15.15.7) that L is finite-modular and (31.1.5) holds.

Finally we shall show that L is irreducible. For any two different atoms Kx and Ky, it is easily seen that $K(x+y)$ is the third atom contained in $Kx \vee Ky$. Hence by (11.6.1) any two atoms are perspective. By (6.7.1) a non-zero central element Z of L contains all atoms and hence $Z = E$. Therefore L is irreducible. ▯

Definition (31.2). Let K_0 be the field of real numbers or that of complex numbers.

A vector space E over K_0 with a topology is called a *topological vector space* (or a *linear topological space*) when

(LT 1) $(x,y)\to x+y$ is continuous on $E\times E$ into E,

and

(LT 2) $(\lambda,x)\to\lambda x$ is continuous on $K_0\times E$ into E.

Let A be a subspace of a topological vector space E. Then A is a topological vector space with the topology induced by E. The quotient space E/A is a topological vector space with the quotient topology. The topology of E/A is Hausdorff if and only if A is closed in E (see Schaefer [1], p. 20).

The set of all closed subspace of a topological vector space E is denoted by $L_c(E)$.

Remark (31.3). Every one-dimensional Hausdorff topological vector space E is isomorphic to K_0, more precisely, for a non-zero element x_0 of E, the mapping $\lambda\to\lambda x_0$ is a homeomorphism of K_0 onto E (see Schaefer [1], p. 21). Hence E is complete.

Lemma (31.4). *Let E be a Hausdorff topological vector space. The set $L_c(E)$ of all closed subspaces of E forms an irreducible complete finite-modular AC-lattice, ordered by set-inclusion.*

Proof. It suffices to show that $L_c(E)$ satisfies the three conditions in (31.1). (1) and (2) are evidently satisfied. Let A be a closed subspace of E and $B=A+K_0x$. We may assume $x\notin A$. The natural mapping φ of E onto the quotient space E/A is continuous and E/A is Hausdorff. Since B/A is a one-dimensional subspace of E/A, it is complete by (31.3), and hence it is closed in E/A. Therefore $B=\varphi^{-1}(B/A)$ is closed in E. ∎

Remark that for $A,B\in L_c(E)$ the meet $A\wedge B$ is the set-intersection $A\cap B$ and the join $A\vee B$ is the closure of the linear sum $A+B$.

Definition (31.5). Let E be a topological vector space. We denote by E' the set of all continuous linear forms on E. Evidently E' is a vector space over K_0.

Lemma (31.6). *Let E be a Hausdorff topological vector space. A closed subspace H is a dual-atom of $L_c(E)$ if and only if there exists a non-zero form $f\in E'$ such that $f^{-1}(0)=H$.*

Proof. (I) If $f \in E'$ and $f \neq 0$, then $H = f^{-1}(0)$ is a closed proper subspace of E. If $H < A$ in $L_c(E)$, then taking $a \in A$ such that $a \notin H$, we have $f(a) \neq 0$. For every $x \in E$, we have

$$x - f(x)\, f(a)^{-1} a \in H,$$

whence $x \in H + K_0 a \leqq A$. Therefore $A = E$, which means that H is a dual-atom of $L_c(E)$.

(II) If H is a dual-atom of $L_c(E)$, then E/H is one-dimensional; because, taking $a \notin H$, since $H + K_0 a \in L_c(E)$, we have $E = H + K_0 a$. We denote by φ the natural mapping of E onto E/H and denote by ψ an isomorphism of E/H onto K_0. Then $f = \psi \circ \varphi$ is a continuous linear form on E such that $H = f^{-1}(0)$. □

Definition (31.7). A Hausdorff topological vector space E is called *locally convex* when for any neighborhood U of 0 there exists a convex neighborhood contained in U (see Schaefer [1], p. 47).

Any locally convex topology on a vector space E is determined by a family $\{p_\alpha; \alpha \in I\}$ of semi-norms on E such that $p_\alpha(x) = 0$ for every $\alpha \in I$ implies $x = 0$. The family of sets

$$\{x \in E; p_{\alpha_i}(x) < \varepsilon \text{ for } \alpha_i, \ldots, \alpha_n \in I \text{ and for } \varepsilon > 0\}$$

forms a base of 0-neighborhoods for this topology (see Schaefer [1], p. 48).

Remark (31.8). The following theorem is a geometrical form of the Hahn-Banach theorem (see Schaefer [1], p. 46).

Theorem. Let E be a topological vector space. If M is a linear manifold in E and if U is a non-empty convex open subset of E, not intersecting M, then there exists a closed hyperplane in E, containing M and not intersecting U.

Lemma (31.9). *Let E be a locally convex space. If A is a closed subspace of E and if* $x \notin A$ *then there exists a dual-atom H of* $L_c(E)$ *such that* $A \leqq H$ *and* $x \notin H$, *and then there exists* $f \in E'$ *such that* $f = 0$ *on A and* $f(x) \neq 0$.

Proof. Since E is locally convex, there exists a convex open neighborhood U of x such that $U \cap A = \emptyset$. By (31.8) there exists a closed hyperplane H such that $A \subset H$ and $U \cap H = \emptyset$. Since $0 \in A \subset H$, H is a subspace and hence it is a dual-atom of $L_c(E)$. Evidently $A \leqq H$ in $L_c(E)$ and $x \notin H$.

By (31.6) there exists $f \in E'$ such that $f^{-1}(0) = H$. Evidently $f = 0$ on A and $f(x) \neq 0$. □

Theorem (31.10). *If E is a locally convex space then the set $L_c(E)$ of all closed subspaces of E forms an irreducible complete DAC-lattice. Moreover,*

(31.10.1) *$(A, B) M^*$ in $L_c(E)$ if and only if $A+B$ is closed.*

Proof. By (31.4), $L_c(E)$ is an irreducible complete finite-modular AC-lattice. If $A<B$ in $L_c(E)$, then there is an element $x \in B$ such that $x \notin A$. By (31.9) there exists a dual-atom H of $L_c(E)$ such that $A \leqq H$ and $x \notin H$. Then $B \nleqq H$. Hence $L_c(E)$ is dual-atomistic, and hence $L_c(E)$ is a DAC-lattice by (27.5). (31.10.1) follows from (31.1.5). ▯

Definition (31.11). Let τ_1 and τ_2 be Hausdorff topologies on a vector space E under each of which E is a topological vector space. We say that τ_1 and τ_2 are *L_c-equivalent* when a subspace of E is τ_1-closed if and only if it is τ_2-closed, that is, $L_c(E, \tau_1) = L_c(E, \tau_2)$.

Theorem (31.12). *Let E be a Hausdorff topological vector space. The lattice $L_c(E)$ of closed subspaces of E is dual-atomistic if and only if the given topology on E is L_c-equivalent to a locally convex topology.*

Proof. If the given topology τ is L_c-equivalent to a locally convex topology, then by (31.10) $L_c(E)$ is dual-atomistic.

Conversely, assume that $L_c(E)$ is dual-atomistic. For every dual-atom H of $L_c(E)$ there exists $f_H \in E'$ such that $H = f_H^{-1}(0)$ by (31.6). Putting $p_H(x) = |f_H(x)|$, p_H is a continuous semi-norm on E. If $x \neq 0$ in E, then since $L_c(E)$ is dual-atomistic, there exists a dual-atom H such that $K_0 x \nleqq H$. Since $f_H(x) \neq 0$, we have $p_H(x) = 0$. Therefore, the family $\{p_H\}$, where H runs through all dual-atoms, determines a locally convex topology σ on E. Since every p_H is τ-continuous, τ is finer than or equal to σ. Hence every σ-closed subspace is τ-closed. Conversely, if A is a τ-closed subspace of E, then since $L_c(E)$ is dual-atomistic, A is the meet of a family $\{H_\alpha\}$ of dual-atoms. Since every H_α is σ-closed and since A is their set-intersection, A is σ-closed. Therefore τ and σ are L_c-equivalent. ▯

Remark (31.13). There is a Hausdorff topological vector space E on which there exists no non-zero continuous linear form (see Schaefer [1], pp. 29—30). In this case the lattice $L_c(E)$ has no dual-atom by (31.6).

32. Modular Pairs in the Lattice of Closed Subspaces

If E is a locally convex space, then by (31.10) the lattice $L_c(E)$ of closed subspace of E is a DAC-lattice, and hence $L_c(E)$ is M-symmetric and M*-symmetric by (27.6). Moreover, $(A, B) M^*$ in $L_c(E)$ if and only if $A+B$ is closed. In this section, we shall give some conditions equivalent to $(A,B)M$ in $L_c(E)$.

Definition (32.1). Let E be a locally convex space and let E' be the vector space of continuous linear forms on E. The *weak topology* $\sigma(E,E')$ is the coarsest topology on E for which each form $f\in E'$ is continuous.

For any $x\in E$, the mapping $f\to f(x)$ of E' into K_0 is a linear form on E'. The *weak topology* $\sigma(E',E)$ is the coarsest topology on E' for which each form $f\to f(x)$ on E' is continuous.

These weak topologies are locally convex topologies on E and E' respectively. The locally convex space E (resp. E') with the topology $\sigma(E,E')$ (resp. $\sigma(E',E)$) is denoted by E_σ (resp. E'_σ).

The dual $(E_\sigma)'$ coincides with E'; and the dual $(E'_\sigma)'$ coincides with E, namely every continuous linear form on E'_σ has the form $f\to f(x)$ for some $x\in E$ (see Schaefer [1], p. 124).

Lemma (32.2). *Let E be a locally convex space. The weak topology $\sigma(E,E')$ is L_c-equivalent to the given topology on E, that is, $L_c(E)=L_c(E_\sigma)$.*

Proof. Since $\sigma(E,E')$ is coarser than or equal to the given topology, any $\sigma(E,E')$-closed subspace is closed. Conversely, if A is a closed subspace, then, by (31.9), for every $x\notin A$ there exists $f\in E'$ such that $f=0$ on A and $f(x)\neq 0$. Hence x does not belong to the $\sigma(E,E')$-closure of A. Therefore A is $\sigma(E,E')$-closed.

Definition (32.3). Let E be a locally convex space. For any subspace A of E, we denote by A^0 the set $\{f\in E'; f=0 \text{ on } A\}$. For any subspace A' of E', we denote by A'^0 the set $\{x\in E; f(x)=0 \text{ for every } f\in A'\}$.

Lemma (32.4). *Let E be a locally convex space.*

(32.4.1) *For any subspace A of E, A^0 is a closed subspace of E'_σ. For any subspace A' of E', A'^0 is a closed subspace of E.*

(32.4.2) *A subspace A of E is closed if and only if $A=A^{00}$. A subspace A' of E'_σ is closed if and only if $A'=A'^{00}$.*

(32.4.3) *$L_c(E)$ is dual-isomorphic to $L_c(E'_\sigma)$ by the mapping $A\to A^0$.*

Proof. (I) Evidently A^0 is a subspace of E'_σ. For $x\in E$, the set $(K_0x)^0 =\{f\in E'; f(x)=0\}$ is closed in E'_σ, since the form $f\to f(x)$ is continuous on E'_σ. Hence $A^0=\bigcap((K_0x)^0;\, x\in A)$ is closed in E'_σ.

Similarly, A'^0 is a closed subspace of E_σ, and hence it is closed in E by (32.2).

(II) A^{00} (resp. A'^{00}) is a closed subspace of E (resp. E'_σ) by (I).

Let A be a closed subspace of E. Evidently $A\subset A^{00}$. If $x\notin A$ then by (31.9) there exists $f\in E'$ such that $f\in A^0$ and $f(x)\neq 0$. Hence $x\notin A^{00}$. Therefore $A=A^{00}$. Similarly, $A'=A'^{00}$ when A' is a closed subspace of E'_σ.

(III) It follows from (I) and (II) that $A \to A^0$ and $A' \to A'^0$ are mutually inverse mappings between $L_c(E)$ and $L_c(E'_\sigma)$. Since $A \subset B$ implies $A^0 \supset B^0$, the mapping $A \to A^0$ is a dual-isomorphism. □

Theorem (32.5). *Let A and B be closed subspaces of a locally convex space E. The following four statements are equivalent.*

(α) $(A, B)M$ *in* $L_c(E)$.
(β) $A^0 + B^0 = (A \wedge B)^0$.
(γ) *If* $f_1, f_2 \in E'$ *and* $f_1 = f_2$ *on* $A \wedge B$ *then there exists* $f \in E'$ *such that* $f = f_1$ *on A and* $f = f_2$ *on B.*
(δ) *If g is a continuous linear form on the product space* $A \times B$ *such that* $g = 0$ *on the set* $\{(x,y) \in A \times B; x + y = 0\}$ *then there exists* $f \in E'$ *such that* $g(x,y) = f(x+y)$ *for every* $(x,y) \in A \times B$.

Proof. (α)$\Leftrightarrow$(β). By (32.4.3), the statement (α) is equivalent to $(A^0, B^0)M^*$ in $L_c(E'_\sigma)$. This is equivalent to $A^0 + B^0 \in L_c(E'_\sigma)$ by (31.10.1). If (β) holds then $A^0 + B^0 \in L_c(E'_\sigma)$ by (32.4.1). Conversely, if $A^0 + B^0 \in L_c(E'_\sigma)$, then $A^0 + B^0 = A^0 \vee B^0 = (A \wedge B)^0$.

(β)$\Rightarrow$(γ). Assume that (β) holds. Let $f_1, f_2 \in E'$ and $f_1 = f_2$ on $A \wedge B$. Since $f_1 - f_2 \in (A \wedge B)^0 = A^0 + B^0$, there exist $f_3 \in A^0$ and $f_4 \in B^0$ such that $f_1 - f_2 = f_3 + f_4$. We put $f = f_1 - f_3 = f_2 + f_4$. Then $f = f_1$ on A since $f_3 \in A^0$, and $f = f_2$ on B since $f_4 \in B^0$.

(γ)$\Rightarrow$(δ). Let g be a continuous linear form on $A \times B$ such that $g(x,y) = 0$ whenever $x + y = 0$. We put $f_1(x) = g(x,0)$ for $x \in A$ and put $f_2(y) = g(0,y)$ for $y \in B$. Then f_1 (resp. f_2) is a continuous linear form on A (resp. B). By the Hahn-Banach theorem each f_i can be extended as an element of E' (see Schaefer [1], p. 49). If $x \in A \wedge B$, then since

$$g(x,0) - g(0,x) = g(x, -x) = 0,$$

we have $f_1(x) = f_2(x)$. Hence by (γ) there exists $f \in E'$ such that $f = f_1$ on A and $f = f_2$ on B. For $x \in A$ and $y \in B$, we have

$$f(x+y) = f(x) + f(y) = f_1(x) + f_2(y) = g(x,0) + g(0,y) = g(x,y).$$

(δ)$\Rightarrow$(β). It is evident that $A^0 + B^0 \subset (A \wedge B)^0$. Let $f \in (A \wedge B)^0$. Putting $g(x,y) = f(x)$ for $(x,y) \in A \times B$, g is a continuous linear form on $A \times B$. If $x + y = 0$, then $x = -y \in A \wedge B$, whence $g(x,y) = f(x) = 0$. Hence by (δ) there exists $f_1 \in E'$ such that

$$f_1(x+y) = g(x,y) \quad \text{for every } (x,y) \in A \times B.$$

Then for $x \in A$ we have

$$f_1(x) = g(x,0) = f(x),$$

and for $y \in B$ we have

$$f_1(y) = g(0, y) = f(0) = 0.$$

Hence $f = (f - f_1) + f_1 \in A^0 + B^0$. □

Remark (32.6). Let A be a closed subspace of a locally convex space E. Evidently A and E/A are locally convex spaces. The weak topology on A coincides with the topology induced by the weak topology $\sigma(E, E')$, and the weak topology on E/A coincides with the quotient topology of $\sigma(E, E')$ (see Schaefer [1], p. 135).

Definition (32.7). Let E and F be topological vector spaces. A linear mapping φ of E into F is called a *topological homomorphism* (or briefly, a *homomorphism*) when φ is continuous and open (the image of any open subset of E is open in $\varphi(E)$) (see Schaefer [1], p. 75).

Remark (32.8). Let E and F be locally convex spaces and let N be a closed subspace of E. If φ is a linear mapping of E into F such that $\varphi = 0$ on N, then there exists a mapping φ_0 of E/N into F associated with φ. Since the canonical mapping of E onto E/N is a topological homomorphism, it is easily seen that φ is continuous (resp. open) if and only if φ_0 is continuous (resp. open).

Lemma (32.9). *Let E and F be locally convex spaces. A continuous mapping φ of E into F is a topological homomorphism of E_σ into F_σ if and only if*

(32.9.1) *for every $g \in E'$ such that $g = 0$ on $\varphi^{-1}(0)$ there exists $f \in F'$ such that $g = f \circ \varphi$.*

Proof. (I) It is easy to show that φ is a continuous mapping of E_σ into F_σ (see Schaefer [1], p. 158). We put $N = \varphi^{-1}(0)$, which is a closed subspace of E_σ.

Assume that φ is an open mapping of E_σ into F_σ. By (32.8) the associated one-to-one mapping φ_0 of E_σ/N into F_σ is open. If $g \in E'$ and $g = 0$ on N, then since g is a continuous linear form on E_σ, the associated linear form g_0 on E_σ/N is continuous. Hence the composition $f = g_0 \circ \varphi_0^{-1}$ is a continuous linear form on the subspace $\varphi(E)$ of F_σ. By the Hahn-Banach theorem f can be extended as a continuous linear form on F_σ. Hence $f \in F'$. For $x \in E$ we have

$$f(\varphi(x)) = g_0(\varphi_0^{-1}(\varphi(x))) = g_0(x + N) = g(x).$$

(II) Conversely, assume that (32.9.1) holds. For any continuous linear form g_0 on E_σ/N, putting $g(x) = g_0(x + N)$, g is a continuous linear form on E_σ and hence $g \in E'$. Moreover, $g = 0$ on N. By (32.9.1) there exists $f \in F'$ such that $g = f \circ \varphi$. Then we have $g_0 = f \circ \varphi_0$. Hence for any 0-neighborhood U in E_σ/N such that

$$U = \{x + N; |g_0^{(i)}(x + N)| < \varepsilon \text{ for } i = 1, \ldots, n\},$$

we have $\varphi_0(U) = \{\varphi(x); |f^{(i)}(x)| < \varepsilon \text{ for } i = 1, \ldots, n\}$, which is a 0-neighborhood in the subspace $\varphi(E)$ of F_σ. Hence φ_0 is open, and so is φ. □

Theorem (32.10). *Let A and B be closed subspaces of a locally convex space E. $(A, B)M$ in $L_c(E)$ if and only if*

(32.10.1) *the mapping $\varphi: (x, y) \to x + y$ of $(A \times B)_\sigma$ into E_σ is a topological homomorphism* (in other words, φ is a *weak homomorphism* of $A \times B$ into E).

Proof. φ is a continuous mapping of $A \times B$ into E by (LT 1) in (31.2). Hence it follows from (32.9) that (32.10.1) is equivalent to the statement (δ) in (32.5). □

Corollary (32.11). *Let A and B be closed subspaces of a locally convex space E. $(A, B)M$ in $L_c(E)$ if*

(32.11.1) *the mapping $\varphi: (x, y) \to x + y$ of $A \times B$ into E is a topological homomorphism.*

If E is metrizable, then $(A, B)M$ in $L_c(E)$ is equivalent to (32.11.1).

Proof. Since any topological homomorphism of a locally convex space into a locally convex space is a weak homomorphism (see Schaefer [1], p. 159), (32.11.1) implies (32.10.1).

If E is metrizable then both the domain and the range of φ are metrizable and hence they are Mackey spaces (see Schaefer [1], p. 132). Hence (32.10.1) implies (32.11.1) (see Schaefer [1], p. 159). □

Definition (32.12). A complete metrizable locally convex space is called a *Fréchet space.* Evidently every closed subspace of a Fréchet space is also a Fréchet space.

Remark (32.13). By Banach's homomorphism theorem we obtain the following result (see Schaefer [1], p. 77).

Let E and F be Fréchet spaces and let φ be a continuous linear mapping of E into F. Then φ is a topological homomorphism if and only if $\varphi(E)$ is closed in F.

Theorem (32.14). *Let E be a Fréchet space and let A and B be closed subspaces of E. The following three statements are equivalent.*

(α) *$(A, B)M$ in $L_c(E)$.*

(β) *$(A, B)M^*$ in $L_c(E)$.*

(γ) *The mapping $\varphi: (x, y) \to x + y$ of $A \times B$ into E is a topological homomorphism.*

Proof. The equivalence of (α) and (γ) follows from (32.11). It follows from (32.13) that (γ) holds if and only if $\varphi(A\times B)=A+B$ is closed in E. Hence (γ) is equivalent to (β) by (31.10.1). □

Remark (32.15). If E is a Fréchet space then by (32.14) $L_c(E)$ satisfies all the statements in (27.15).

On the other hand, there is a non-complete normed space E such that $L_c(E)$ satisfies none of them. Let F be a non-complete normed space and let $\hat{F}$ be its completion. Taking $x\in\hat{F}$ such that $x\notin F$, we consider the following subspace of $\hat{F}\times\hat{F}$:

$$E=F\times F+K_0(x,x).$$

Evidently, E is a non-complete normed space. We put

$$A=F\times\{0\},\qquad B=\{0\}\times F\quad\text{and}\quad C=B+K_0(x,x).$$

It is easily seen that A, B and C are closed subspaces of E. We shall show that in $L_c(E)$

(1) the pair (A,B) is modular but is not dual-modular,

(2) the pair (A,C) is dual-modular but is not modular.

Since $A+B=F\times F$, we have $(x,x)\in\overline{A+B}$, whence $A+B$ is not closed in E. Hence (A,B) is not dual-modular. But (A,C) is dual-modular since $A+C=E$. Since $A+B=F\times F$ is isomorphic to $A\times B$, (A,B) is modular by (32.11). Since $(A,C)M^*$ and $B\lessdot C$, if $(A,C)M$ then by (27.14) we would have $(A,B)M^*$, a contradiction. Hence (A,C) is not modular.

Remark (32.16). Let E be a metrizable locally convex space. If E is separable (that is, E has a countable dense subset) then E'_σ is also separable (see Schaefer [1], p. 128).

Theorem (32.17). *Let E be a normed space. E is finite-dimensional if and only if the lattice* $L_c(E)$ *is modular.*

Proof. (I) If E is finite-dimensional, then $L_c(E)$ is modular since $L_c(E)=L(E)$.

(II) Let E be infinite-dimensional. We need to show that $L_c(E)$ is not modular. Evidently E has an infinite-dimensional closed subspace E_1 which is separable. Since $L_c(E_1)$ is a sublattice of $L_c(E)$, it suffices to show that $L_c(E_1)$ is not modular. Hence we may assume that E is separable.

Since E'_σ is separable by (32.16), there exists a sequence $\{A'_n\}$ of members of $L(E'_\sigma)$ such that

$$\{0\}=A'_0\lessdot A'_1\lessdot A'_2\lessdot\cdots\quad\text{and}\quad\bigvee_n A'_n=E'.$$

Putting $A_n = A_n'^0$, by (32.4) we have

$$E = A_0 > A_1 > A_2 > \cdots \quad \text{and} \quad \bigwedge_n A_n = \{0\}.$$

For any $k = 1, 2, \ldots$, there exists a two-dimensional subspace B_k such that $A_{2k} \wedge B_k = \{0\}$ and $A_{2k} + B_k = A_{2k-2}$. Since B_k is a two-dimensional normed space, there exist two linearly independent vectors x_k and y_k such that $\|x_k\| = 1$. Replacing y_k by $x_k + \varepsilon y_k$ for sufficiently small $\varepsilon > 0$, we may assume that $\|x_k - y_k\| < \dfrac{1}{k}$.

(III) We denote by A and B the closed subspaces generated by $\{x_k; k = 1, 2, \ldots\}$ and by $\{y_k; k = 1, 2, \ldots\}$ respectively. We shall show that $A \wedge B = \{0\}$. For any k we have

$$A \leqq K_0 x_1 + \cdots + K_0 x_k + A_{2k},$$

since $x_n \in B_n \leqq A_{2k}$ for every $n > k$. Similarly

$$B \leqq K_0 y_1 + \cdots + K_0 y_k + A_{2k}.$$

Since $x_1, \ldots, x_k$, $y_1, \ldots, y_k$ are linearly independent and since $(B_1 + \cdots + B_k) \wedge A_{2k} = \{0\}$, it is easily seen that $A \wedge B \leqq A_{2k}$. Since $\bigwedge_k A_{2k} = \{0\}$, we have $A \wedge B = \{0\}$.

The one-to-one mapping $\varphi(x, y) = x + y$ of $A \times B$ into E is not an isomorphism, since $\varphi(x_k, -y_k) = x_k - y_k$ converges to 0 while $(x_k, -y_k)$ does not converge to $(0, 0)$. Hence the pair (A, B) is not modular in $L_c(E)$ by (32.11). □

Remark (32.18). We can find some examples of an infinite-dimensional locally convex space E such that $L_c(E)$ is modular.

(I) For any infinite-dimensional vector space E there exists the finest locally convex topology on E (see Schaefer [1], p. 56). When E has this topology, $L_c(E)$ coincides with $L(E)$, whence $L_c(E)$ is modular (see Schaefer [1], p. 69, Exercise 7).

(II) Let E be a locally convex space. E is said to be *minimal* if there exists no strictly coarser locally convex topology (Hausdorff) on E.

It can be proved (see Schaefer [1], p. 191, Exercise 6) that E is minimal if and only if E is isomorphic to the product space of a family of one-dimensional locally convex spaces and that if E is minimal then $L_c(E)$ is modular. Hence the product space E_0 of an infinite sequence of one-dimensional locally convex spaces is an infinite-dimensional Fréchet space such that $L_c(E_0)$ is modular.

Remark (32.19). Let E be an infinite-dimensional metrizable locally convex space which is separable. In the same way as in the proof of (32.17) we can prove that $L_c(E)$ is not modular if E satisfies the following condition:

(32.19.1) If $\{A_n\}$ is a sequence of closed subspaces of E such that $E=A_0>A_1>A_2>\cdots$ and $\bigwedge_n A_n=\{0\}$, then there exists a neighborhood U of 0 such that $U \not\supset A_n$ for every n.

Using this fact, we can prove that a Fréchet space E is minimal if and only if $L_c(E)$ is modular (see Martineau [1]).

EXERCISE 32.1. Prove that if an infinite-dimensional metrizable locally convex space E is separable and satisfies (32.19.1) then $L_c(E)$ is not modular.

33. Pairs of Dual Spaces

Definition (33.1). Let E be a left vector space over a division ring K, and let F be a right vector space over K. The pair (E,F) is called a *pair of dual spaces* if there exists a mapping f of $E\times F$ onto K satisfying the following four conditions; where $x,x_i\in E$, $y,y_i\in F$, $\lambda,\mu\in K$.

(PD 1) $\quad f(\lambda x_1+\mu x_2,y)=\lambda f(x_1,y)+\mu f(x_2,y)$.

(PD 2) $\quad f(x,y_1\lambda+y_2\mu)=f(x,y_1)\lambda+f(x,y_2)\mu$.

(PD 3) $\quad f(x,y)=0$ for every y implies $x=0$.

(PD 4) $\quad f(x,y)=0$ for every x implies $y=0$.

The set E^* of all linear forms on E forms a right vector space over K by the following conventions: For $g,g_i\in E^*$

$$(g_1+g_2)(x)=g_1(x)+g_2(x) \quad \text{and} \quad (g\lambda)(x)=g(x)\lambda.$$

For any $y\in F$, we can define a form $f_y\in E^*$ by putting $f_y(x)=f(x,y)$. By (PD 2) and (PD 4), the mapping $y\to f_y$ is an isomorphism of F into E^*. Similarly, putting $f_x(y)=f(x,y)$, the mapping $x\to f_x$ is an isomorphism of E into the left vector space F^* of all linear forms on F.

Definition (33.2). Let (E,F) be a pair of dual spaces with f. For a subspace A of E we put

$$A^0=\{y\in F;\, f(x,y)=0 \text{ for every } x\in A\},$$

and for a subspace B of F we put

$$B^0=\{x\in E;\, f(x,y)=0 \text{ for every } y\in B\}.$$

By (PD 1) and (PD 2), A^0 and B^0 are subspaces of F and E respectively, and we have $A\subset A^{00}$ and $B\subset B^{00}$. A subspace A of E is called F-*closed* if $A^{00}=A$, and a subspace B of F is called E-*closed* if $B^{00}=B$. By (PD 3) and (PD 4), both E and $\{0_E\}$ are F-closed and both F and $\{0_F\}$

are E-closed. Moreover, A^0 is an E-closed subspace of F and B^0 is an F-closed subspace of E.

Lemma (33.3). *Let* (E,F) *be a pair of dual spaces.*

(33.3.1) *If* A_α *is an F-closed subspace of E for every* $\alpha \in I$ *then the set-intersection* $\bigcap(A_\alpha; \alpha \in I)$ *is also F-closed.*

(33.3.2) *If A is an F-closed subspace of E, then* $A+Ka$ *is F-closed for every* $a \in E$.

The corresponding statements for E-closed subspaces of F also hold.

Proof. (I) If the spaces A_α are F-closed, then since $\bigcap_\alpha A_\alpha \subset A_\beta$ for every $\beta \in I$, we have

$$(\bigcap_\alpha A_\alpha)^{00} \subset A_\beta^{00} = A_\beta .$$

Hence $(\bigcap_\alpha A_\alpha)^{00} \subset \bigcap_\alpha A_\alpha$, which implies that $\bigcap_\alpha A_\alpha$ is F-closed.

(II) To prove (33.3.2), we may assume that $a \notin A$. Since $a \notin A^{00}$, there exists $b \in A^0$ such that $f(a,b) \neq 0$. We may assume $f(a,b)=1$, replacing b by $bf(a,b)^{-1}$. Then for every $y \in A^0$ we have

$$y - bf(a,y) \in (A+Ka)^0,$$

because $y-bf(a,y) \in A^0$ and

$$f(a,y-bf(a,y)) = f(a,y) - f(a,b)f(a,y) = 0.$$

Let $x \in (A+Ka)^{00}$. Then for every $y \in A^0$ we have

$$0 = f(x,y-bf(a,y)) = f(x,y) - f(x,b)f(a,y) = f(x-f(x,b)a,y).$$

Hence $x - f(x,b)a \in A^{00} = A$, whence $x \in A+Ka$. Therefore $A+Ka$ is F-closed. ▯

Theorem (33.4). *Let* (E,F) *be a pair of dual spaces. Both the set* $L_F(E)$ *of all F-closed subspaces of E and the set* $L_E(F)$ *of all E-closed subspaces of F form irreducible complete DAC-lattices, ordered by set-inclusion. Moreover, for* $A, B \in L_F(E)$,

(33.4.1) $(A,B)M^*$ *in* $L_F(E)$ *if and only if* $A+B$ *is F-closed,*

(33.4.2) $(A,B)M$ *in* $L_F(E)$ *if and only if* A^0+B^0 *is E-closed* (or equivalently, $A^0+B^0=(A \wedge B)^0$).

Proof. By (33.3.1) and (33.3.2), $L_F(E)$ satisfies the three conditions in (31.1). Hence $L_F(E)$ forms an irreducible complete AC-lattice, ordered by set-inclusion, and (33.4.1) holds. Similarly, $L_E(F)$ forms an irreducible complete AC-lattice, ordered by set-inclusion. Since two lattices $L_F(E)$ and $L_E(F)$ are dual-isomorphic by the mapping $A \to A^0$, they are DAC-lattices. Moreover, $(A,B)M$ in $L_F(E)$ is equivalent to $(A^0,B^0)M^*$ in $L_E(F)$. Hence (33.4.2) holds. ▯

Remark (33.5). In the definition (33.1) of a pair (E,F) of dual spaces over a division ring K, if K is commutative, then F may be a left vector space.

If E is a locally convex space, then evidently (E,E') forms a pair of dual spaces over K_0, and then by (32.4.2) we have

$$L_{E'}(E)=L_c(E) \quad \text{and} \quad L_E(E')=L_c(E'_\sigma).$$

Remark (33.6). A representation theorem of irreducible complete DAC-lattices is based on the following representation theorem.

Theorem. If L is an irreducible modular matroid lattice of length $\geqq 4$, then there exists a vector space E over a division ring K such that L is isomorphic to the lattice $L(E)$ of all subspaces of E.

An outline of the proof of this theorem, given in Chapter VII of Baer [1], is as follows:

(1) Let h be a fixed dual-atom of L. We denote by Γ the group of automorphisms σ of L such that $\sigma(x)=x$ for every $x\in L[0,h]$. It can be proved that for any $\sigma\in\Gamma$ with $\sigma\neq 1$, there exists one and only one atom p such that

$$x\vee p=\sigma(x)\vee p \quad \text{for every } x\in L.$$

We denote this unique atom by $C(\sigma)$.

(2) Using the assumption that L is of length $\geqq 4$, we can prove that the theorem of Desargues holds in L. By the aid of this theorem we can prove the following statements (3) and (4).

(3) If p and q are different atoms such that $p\nleqq h$ and $q\nleqq h$ and if r is a third atom contained in $p\vee q$, then there exists one and only one automorphism $\sigma\in\Gamma$ such that $\sigma(p)=q$ and $C(\sigma)=r$.

(4) If $\sigma_1,\sigma_2\in\Gamma$ and each of σ_1, σ_2 and $\sigma_2\sigma_1$ is not the identity then $C(\sigma_2\sigma_1)\leqq C(\sigma_1)\vee C(\sigma_2)$.

(5) For $a\in L[0,h]$ we put

$$\Gamma(a)=\{\sigma\in\Gamma;\sigma=1 \text{ or } C(\sigma)\leqq a\}.$$

By (4), $\Gamma(a)$ is a subgroup of Γ for every $a\in L[0,h]$. Especially, it can be proved that $\Gamma(h)$ is a normal subgroup, and moreover, using (3), it is commutative.

(6) Let $\beta\to\beta^\eta$ be a non-trivial endomorphism of the commutative group $\Gamma(h)$. It can be proved that $\Gamma(a)^\eta\subset\Gamma(a)$ for every $a\in L[0,h]$ if and only if there exists an element $\sigma\in\Gamma$ such that

$$\beta^\eta=\sigma^{-1}\beta\sigma \quad \text{for every } \beta\in\Gamma(h).$$

From this, the set of endomorphisms η of $\Gamma(h)$ such that $\Gamma(a)^\eta\subset\Gamma(a)$ for every $a\in L[0,h]$ forms a division ring K, and $\Gamma(h)$ can be considered as a vector space over K. Then $\Gamma(a)$ is a subspace of $\Gamma(h)$ for every $a\in L[0,h]$.

(The above results (1)~(6) are valid when L is an irreducible complete DAC-lattice of length $\geqq 4$.)

(7) It can be proved by (3) that the order-preserving mapping $a \to \Gamma(a)$ of $L[0,h]$ into the lattice $L(\Gamma(h))$ of all subspaces of $\Gamma(h)$ is one-to-one. Moreover, we can prove that this mapping is onto, using the fact that L is compactly atomistic. Hence $L[0,h]$ and $L(\Gamma(h))$ are isomorphic by this mapping.

(8) Using the theorem of Desargues, we can construct an irreducible modular matroid lattice $\bar{L}$ such that L is isomorphic to the sublattice $\bar{L}[0,\bar{h}]$ where $\bar{h}$ is a dual-atom of $\bar{L}$. Applying the results (1)~(7) to $\bar{L}$ and $\bar{h}$, we obtain the vector space $\Gamma(\bar{h})$ such that $L(\Gamma(\bar{h})) \cong \bar{L}[0,\bar{h}] \cong L$.

The last step (8) can be replaced by the following one (see McLaughlin [1]).

(8′) Let E be the product space of $\Gamma(h)$ and a one-dimensional vector space over K. In a manner similar to (3) we can prove that L is isomorphic to $L(E)$.

Theorem (33.7). *Let L be an irreducible complete DAC-lattice of length $\geqq 4$. There exists a pair of dual spaces (E,F) over a division ring K such that L is isomorphic to the lattice $L_F(E)$ of F-closed subspaces of E.*

Proof. (I) Since the lattice $L(\Omega(L))$ of subspaces of $\Omega(L)$ is a modular matroid lattice of length $\geqq 4$ by (27.16), and since it is irreducible by (16.7), it follows from (33.6) that there exists a left vector space E over a division ring K such that $L(E) \cong L(\Omega(L))$. By (27.16) and (27.17) there is a one-to-one order-preserving mapping ω of L into $L(E)$ with the following properties:

(1) $$\omega(0)=\{0\} \quad \text{and} \quad \omega(1)=E,$$

(2) $$\omega(\bigwedge_\alpha a_\alpha)=\bigwedge_\alpha \omega(a_\alpha),$$

(3) $$\omega(\mathscr{I}(L)) = \mathscr{I}(L(E)).$$

(4) for any dual-atom h of L, $\omega(h)$ is a dual-atom of $L(E)$.

(5) $$\omega(\mathscr{I}(L^*)) \text{ is an ideal of } \mathscr{I}(L(E)^*).$$

Let E^* be the right vector space of all linear forms on E. For every dual-atom h of L, since $\omega(h)$ is a dual-atom of $L(E)$ by (4), there exists $y \in E^*$ such that

$$\omega(h)=\{x \in E;\, y(x)=0\}=\ker(y).$$

(II) Let F be the set of all $y \in E^*$ such that $\ker(y)=\omega(h)$ for some dual-atom h of L, with $0 \in E^*$ added. We shall show that F is a subspace of E^*. Evidently, if $y \in F$ then $y\lambda \in F$ for every $\lambda \in K$. Let $y_1, y_2 \in F$. To prove $y_1+y_2 \in F$, we may assume that $y_1 \neq 0$, $y_2 \neq 0$ and $y_1+y_2 \neq 0$.

Then there are dual-atoms h_1 and h_2 such that $\omega(h_i)=\ker(y_i)$ $(i=1,2)$. Since $\ker(y_1+y_2)$ is a dual-atom of $L(E)$ and since

$$\ker(y_1+y_2)\geqq\ker(y_1)\wedge\ker(y_2)=\omega(h_1)\wedge\omega(h_2),$$

it follows from (5) that $\ker(y_1+y_2)\in\omega(\mathcal{J}(L^*))$. Hence there exists a dual-atom h of L such that $\omega(h)=\ker(y_1+y_2)$. Therefore $y_1+y_2\in F$.

(III) Putting $f(x,y)=y(x)$, f is a mapping of $E\times F$ into K satisfying (PD 1), (PD 2) and (PD 4). We shall show that f satisfies (PD 3). If $x\neq 0$, then by (3) there exists an atom p of L such that $\omega(p)=Kx$. Since L is dual-atomistic, there exists a dual-atom h of L with $p\wedge h=0$. Taking $y\in F$ such that $\ker(y)=\omega(h)$, we have

$$Kx\cap\ker(y)=\omega(p)\wedge\omega(h)=\omega(p\wedge h)=\{0\}.$$

Hence $x\notin\ker(y)$, whence $f(x,y)\neq 0$. Therefore f satisfies (PD 3), and then (E,F) is a pair of dual spaces. We remark that for $y\in F$ we have

$$\ker(y)=\{x\in E; f(x,y)=0\}=(yK)^0.$$

(IV) If h is a dual-atom of L, then $\omega(h)$ is an F-closed subspace of E since $\omega(h)=(yK)^0$ for some $y\in F$. For any $a\in L$, since $a=\bigwedge(h; h\geqq a)$, by (2) we have

$$\omega(a)=\bigwedge(\omega(h); h\geqq a).$$

Since the meet in $L(E)$ coincides with set-intersection, $\omega(a)$ is F-closed by (33.3.1).

(V) Let A be an F-closed subspace of E. For $y\in F$ with $y\neq 0$, we denote the dual-atom h satisfying $\omega(h)=(yK)^0$ by h_y. We put $a=\bigwedge(h_y; 0\neq y\in A^0)$. Then

$$\omega(a)=\bigwedge(\omega(h_y); 0\neq y\in A^0)=\bigwedge((yK)^0; 0\neq y\in A^0)=A^{00}=A.$$

(VI) By (IV) and (V), the range of the mapping ω coincides with $L_F(E)$. Since ω is one-to-one and order-preserving, L and $L_F(E)$ are isomorphic by ω. □

EXERCISE 33.1. Let (E,F) be a pair of dual spaces, and let A and B be F-closed subspaces of E. Prove that $(A,B)M$ in $L_F(E)$ if and only if the following condition is satisfied:

If $y_1, y_2\in F$ and $f_{y_1}=f_{y_2}$ on $A\wedge B$ then there exists $y\in F$ such that $f_y=f_{y_1}$ on A and $f_y=f_{y_2}$ on B.

34. Vector Spaces with Hermitian Forms

Definition (34.1). Let K be a division ring with an involutorial anti-automorphism $\lambda\to\lambda^*$ (which means $(\lambda+\mu)^*=\lambda^*+\mu^*$, $(\lambda\mu)^*=\mu^*\lambda^*$ and $\lambda^{**}=\lambda$), and let (E,F) be a pair of dual spaces over K. In this case,

F may be a left vector space, replacing $y\lambda$ by $\lambda^* y$, and then (PD 2) is replaced by

(PD 2*) $$f(x, \lambda y_1 + \mu y_2) = f(x, y_1)\lambda^* + f(x, y_2)\mu^*.$$

A form f on $E \times F$ satisfying (PD 1) and (PD 2*) is called a *semi-bilinear form*.

In the case that $E = F$, a semi-bilinear form f on $E \times E$ is called an *Hermitian form* if f satisfies the following two conditions:

(H 1) $$f(y, x) = f(x, y)^* \quad \text{and}$$

(H 2) $$f(x, x) = 0 \quad \text{implies } x = 0.$$

Theorem (34.2). *Let K be a division ring with an involutorial anti-automorphism. If a vector space E over K has an Hermitian form f, then (E, E) forms a pair of dual spaces, and then the lattice $L_E(E)$ is an irreducible complete orthocomplemented AC-lattice.*

Proof. (E, E) forms a pair of dual spaces, since f satisfies (PD 3) and (PD 4) by (H 2). Hence $L_E(E)$ is an irreducible complete DAC-lattice by (33.4). Since $f(x, y) = 0$ is equivalent to $f(y, x) = 0$ by (H 1), $A \to A^0$ is a mapping of $L_E(E)$ onto itself satisfying (29.1.2) and (29.1.3). Moreover, we have $A \wedge A^0 = \{0\}$ by (H 2). Hence this mapping is an orthocomplementation by (29.2). □

Remark (34.3). A representation theorem of irreducible complete orthocomplemented AC-lattices is based on the theorem in (33.6) and the following theorem.

Theorem. Let E_n be a vector space over a division ring K with finite dimension $n \geqq 4$. If the lattice $L(E_n)$ of subspaces of E_n is orthocomplemented, then there exists an involutorial anti-automorphism $\lambda \to \lambda^*$ of K and there exists an Hermitian form f on $E_n \times E_n$ such that for every $A \in L(E_n)$

(34.3.1) $$A^\perp = \{x \in E_n;\, f(x, y) = 0 \text{ for every } y \in A\}.$$

Moreover, the pair $(*, f)$ is unique in the following sense: If $\lambda \to \lambda^{\bar{*}}$ is an involutorial anti-automorphism and if $\bar{f}$ is an Hermitian form (with respect to $\bar{*}$) satisfying (34.3.1), then there exists $\gamma \in K$ such that

$$\lambda^{\bar{*}} = \gamma^{-1}\lambda^*\gamma \quad \text{for all } \lambda \in K \quad \text{and}$$
$$\bar{f}(x, y) = f(x, y)\gamma \quad \text{for all } x, y \in E_n.$$

An outline of a proof of existence, given in Birkhoff and von Neumann [1], is as follows.

(1) We can take a linearly independent family $\{a_1, \ldots, a_n\}$ of vectors in E_n such that

$$(K a_i)^\perp = K a_1 + \cdots + K a_{i-1} + K a_{i+1} + \cdots + K a_n \qquad (i = 1, \ldots, n).$$

(2) We denote by $H[\mu_1:\ldots:\mu_n]$ the $(n-1)$-dimensional subspace $\{\lambda_1 a_1+\cdots+\lambda_n a_n; \sum_{i=1}^n \lambda_i\mu_i=0\}$, where $(\mu_1,\ldots,\mu_n)\neq(0,\ldots,0)$. We can prove that there exist mappings $\varphi_2,\ldots,\varphi_n$ of K into itself such that if $x=a_1+\lambda_2 a_2+\cdots+\lambda_n a_n$ then

$$(Kx)^\perp = H[1:\varphi_2(\lambda_2):\cdots:\varphi_n(\lambda_n)].$$

(3) Using the fact that $n\geqq 4$, it can be proved that $\varphi_2,\ldots,\varphi_n$ have the following properties.

(a) For each $i=2,\ldots,n$, $\varphi_i(\lambda)=0$ if and only if $\lambda=0$.
(b) $\sum_{i=2}^n \mu_i\varphi_i(\lambda_i)=-1$ if and only if $\sum_{i=2}^n \lambda_i\varphi_i(\mu_i)=-1$.
(c) $\varphi_i(\lambda\mu)=\varphi_i(\mu)\varphi_i(1)^{-1}\varphi_i(\lambda)$ for all $i=2,\ldots,n$.
(d) $\varphi_i(1)^{-1}\varphi_i(\lambda)=\varphi_j(1)^{-1}\varphi_j(\lambda)$ for all $i,j=2,\ldots,n$.

(4) We put $\lambda^*=\varphi_2(1)^{-1}\varphi_2(\lambda)$ for each $\lambda\in K$. Then it can be proved by the properties in (3) that $\lambda\to\lambda^*$ is an involutorial anti-automorphism of K and that

$$\varphi_i(1)^*=\varphi_i(1) \quad \text{for all } i=2,\ldots,n.$$

(5) For two elements $x=\sum_{i=1}^n \lambda_i a_i$ and $y=\sum_{i=1}^n \mu_i a_i$ of E_n, we put

$$f(x,y)=\lambda_1\mu_1^*+\sum_{i=2}^n \lambda_i\varphi_i(1)\mu_i^*.$$

Then it can be proved that f is an Hermitian form on E_n and that f satisfies (34.3.1).

Uniqueness of $(*,f)$ can be proved as follows. If $(\bar{*},\bar{f})$ represents the same orthocomplementation, then $f(x,y)=0$ is equivalent to $\bar{f}(x,y)=0$. Hence we have $\bar{f}(a_i,a_j)=0$ when $i\neq j$ and $\bar{f}(a_1-\varphi_i(1)^{-1}a_i, a_1+a_i)=0$ $(i=2,\ldots,n)$, whence for each such i,

$$\bar{f}(a_1,a_1)=\varphi_i(1)^{-1}\bar{f}(a_i,a_i).$$

Therefore, putting $\gamma=\bar{f}(a_1,a_1)$, we have

$$\bar{f}(a_i,a_i)=f(a_i,a_i)\gamma \quad \text{for every } i=1,\ldots,n.$$

Moreover, since $\bar{f}(a_1-\lambda^*\varphi_2(1)^{-1}a_2, \lambda a_1+a_2)=0$, we have

$$\bar{f}(a_1,a_1)\lambda^{\bar{*}}=\lambda^*\varphi_2(1)^{-1}\bar{f}(a_2,a_2)=\lambda^*\bar{f}(a_1,a_1).$$

Therefore, $\lambda^{\bar{*}}=\gamma^{-1}\lambda^*\gamma$ for every $\lambda\in K$, and

$$\bar{f}(x,y)=\sum_{i=1}^n \lambda_i\bar{f}(a_i,a_i)\mu^{\bar{*}}=\sum_{i=1}^n \lambda_i f(a_i,a_i)\gamma\mu^{\bar{*}}$$
$$=\sum_{i=1}^n \lambda_i f(a_i,a_i)\mu^*\gamma=f(x,y)\gamma.$$

(In Baer [1], Chapter IV and in MacLaren [1], there are precise arguments concerning this theorem, using results on semi-linear transformations.)

Definition (34.4). Let L_1 and L_2 be orthocomplemented lattices, each with orthocomplementation denoted by "$\perp$". We say that L_1 and L_2 are *ortho-isomorphic* when there exists an isomorphism φ of L_1 onto L_2 such that

$$\varphi(a)^{\perp} = \varphi(a^{\perp}) \quad \text{for every } a \in L_1 .$$

Theorem (34.5). *Let L be an irreducible complete orthocomplemented AC-lattice of length $\geqq 4$. There exists a division ring K with an involutorial anti-automorphism $\lambda \to \lambda^*$ and there exists a vector space E over K with an Hermitian form f such that L is ortho-isomorphic to the lattice $L_E(E)$ of E-closed subspaces of E.*

Proof. (I) It follows from (27.16) and (33.6) that there exists a vector space E over a division ring K such that there is a one-to-one order-preserving mapping ω of L into $L(E)$ with the following properties:

(1) $$\omega(0) = \{0\} \quad \text{and} \quad \omega(1) = E.$$

(2) $$\omega(\bigwedge_\alpha a_\alpha) = \bigwedge_\alpha \omega(a_\alpha).$$

(3) $$\omega \text{ is an isomorphism of } \mathscr{I}(L) \text{ onto } \mathscr{I}(L(E)).$$

(II) If L is of finite length n then E is n-dimensional and ω is an isomorphism of L onto $L(E)$ by (3). Hence $L(E)$ has an orthocomplementation induced by that of L. By (34.3), there exists a pair $(*, f)$ such that

$$A^{\perp} = \{x \in E; f(x,y) = 0 \text{ for every } y \in A\} = A^0 .$$

Evidently every subspace of E is E-closed. Hence L is ortho-isomorphic to $L_E(E)$.

(III) Let L be of infinite length. For any $a \in \mathscr{I}(L)$ it follows from (29.10) that $x \to x^{\perp} \wedge a$ is an orthocomplementation in $L[0,a]$. For any finite-dimensional subspace A of E, by (3) there exists $a \in \mathscr{I}(L)$ such that $\omega(a) = A$, and then ω is an isomorphism of $L[0,a]$ onto $L(A)$. Hence $L(A)$ has an orthocomplementation induced by that of $L[0,a]$.

(IV) We take a fixed 4-dimensional subspace A_0 of E. Since $L(A_0)$ is orthocomplemented, by (34.3) there exists an involutorial anti-automorphism $\lambda \to \lambda^*$ of K and there exists an Hermitian form f_0 on $A_0 \times A_0$ such that

$$\omega(b^{\perp}) \wedge A_0 = \{y \in A_0; f_0(x,y) = 0 \text{ for every } x \in B\}$$

for every $B \in L(A_0)$, where $\omega(b) = B$.

For any finite-dimensional subspace A of E with $A_0 \subset A$, there exists a pair $(\bar{*}, f_A)$ such that for every $B \in L(A)$

(4) $$\omega(b^{\perp}) \wedge A = \{y \in A; f_A(x,y) = 0 \text{ for every } x \in B\}$$

where $\omega(b)=B$. By the uniqueness of $(*,f_0)$ on A_0, there exists $\gamma\in K$ such that $\lambda^{\overline{*}}=\gamma^{-1}\lambda^*\gamma$ for all $\lambda\in K$ and $f_A(x,y)=f_0(x,y)\gamma$ for all $x,y\in A_0$. Hence, replacing $(\overline{*},f_A)$ by $(*,f_A\gamma^{-1})$, we have a unique Hermitian form f_A (with respect to $*$) such that f_A induces the orthocomplementation in $L(A)$ and that $f_A=f_0$ on $A_0\times A_0$. It is evident that

(5) $$\text{if } A_0\subset A_1\subset A_2, \quad \text{then } f_{A_1}=f_{A_2} \quad \text{on } A_1\times A_1.$$

(V) For every $x,y\in E$ we put

(6) $$f(x,y)=f_{A_0+Kx+Ky}(x,y).$$

It is easy to show by (5) that f is an Hermitian form on $E\times E$. For every atom p of L, there exists $x\in E$ such that $\omega(p)=Kx$. We shall show that

(7) $$\omega(p^\perp)=\{y\in E; f(x,y)=0\}=(Kx)^0.$$

If $y\in\omega(p^\perp)$, then putting $A=A_0+Kx+Ky$, since $y\in\omega(p^\perp)\wedge A$, we have $f_A(x,y)=0$ by (4). Hence $f(x,y)=0$ by (6). Conversely, if $f(x,y)=0$, then by (6)

$$f_{A_0+Kx+Ky}(x,\lambda y)=0 \quad \text{for every } \lambda\in K,$$

whence $y\in\omega(p^\perp)$ by (4). Therefore (7) holds, and hence $\omega(p^\perp)\in L_E(E)$ for every atom p.

For any $a\in L$, since L is atomistic, we have

$$a=a^{\perp\perp}=(\bigvee(p;p\leqq a^\perp))^\perp=\bigwedge(p^\perp;p\leqq a^\perp).$$

Hence by (2) and (7)

(8) $$\omega(a)=\bigwedge(\omega(p^\perp);p\leqq a^\perp)=\bigcap(\omega(p)^0;p\leqq a^\perp).$$

Therefore $\omega(a)\in L_E(E)$ for every $a\in L$.

(VI) Let $A\in L_E(E)$. We put $a=\bigwedge(p^\perp;\omega(p)\leqq A^0)$. Then

$$\omega(a)=\bigwedge(\omega(p^\perp);\omega(p)\leqq A^0)=\bigwedge(\omega(p)^0;\omega(p)\leqq A^0)=A^{00}=A.$$

(VII) By (V) and (VI), the range of the mapping ω coincides with $L_E(E)$. Since ω is one-to-one and order-preserving, L and $L_E(E)$ are isomorphic by ω. Moreover, they are ortho-isomorphic, since by (8)

$$\omega(a^\perp)=\bigwedge(\omega(p)^0;p\leqq a)=[\bigvee(\omega(p);p\leqq a)]^0=\omega(a)^0 \quad \text{for every } a\in L,$$

where $\bigvee$ is the join in $L_E(E)$. □

Definition (34.6). Let K_0 be the field of real numbers or that of complex numbers. In the former case we put $\lambda^*=\lambda$ and in the latter case we put $\lambda^*=\bar{\lambda}$. A vector space E over K_0 is called an *inner product space* (or a *pre-Hilbert space*) when E has an Hermitian form, which we denote by $\langle x,y\rangle$.

If E is an inner product space, then by (34.2) (E,E) forms a pair of dual spaces and the lattice $L_E(E)$ is an irreducible complete orthocomplemented AC-lattice.

Remark (34.7). Any inner product space E is a normed space with the norm $\|x\| = \sqrt{\langle x,x\rangle}$. Hence the set $L_c(E)$ of all closed subspaces forms an irreducible complet DAC-lattice by (31.10). We remark that $L_E(E)$ does not coincide with $L_c(E)$ in general ($L_E(E)$ is contained in $L_c(E)$).

Theorem (34.8). *Let E be a Hilbert space* (a complete inner product space). *The set $L_c(E)$ of all closed subspaces of E forms an irreducible complete orthomodular AC-lattice which is moreover O-symmetric.*

Proof. Since any closed subspace of E is E-closed (see Halmos [1], p. 24), we have $L_c(E) = L_E(E)$. Hence $L_c(E)$ is an irreducible complete orthocomplemented AC-lattice. For any $A \in L_c(E)$ we have a decomposition $E = A + A^0$ by the projection theorem (see Halmos [1], p. 25). Hence by (33.4.1) we have $(A, A^0)M^*$ in $L_c(E)$. Therefore $L_c(E)$ is orthomodular by (29.13). Moreover, $L_c(E)$ is O-symmetric by (32.14, since E is a Fréchet space. □

Theorem (34.9). *Let E be an inner product space. The lattice $L_E(E)$ of E-closed subspaces of E is orthomodular if and only if E is a Hilbert space.*

Proof. (I) If E is a Hilbert space, then $L_E(E)$ is orthomodular by (34.8). To prove the converse, assume that $L_E(E)$ is orthomodular and let H be the Hilbert space which is the completion of E. For any subset S of H we put

$$S^\perp = \{y \in H; \langle x,y\rangle = 0 \text{ for every } x \in S\}.$$

First, we shall show that if H_1 is a closed subspace of H such that the orthocomplement $H_1^\perp$ is finite-dimensional, then $H_1 \wedge E$ is dense in H_1. It suffices to prove this when $H_1^\perp$ is one-dimensional, that is, $H_1 = \{a\}^\perp$ for some $a \neq 0$. Since E is dense in H, there exists $b \in E$ such that $\langle b,a\rangle \neq 0$.

Let x be an arbitrary element of H_1. For any positive number ε there exists $y \in E$ such that $\|x-y\| < \varepsilon$. Then, since $\langle x,a\rangle = 0$, we have

$$|\langle y,a\rangle| = |\langle y-x,a\rangle| < \varepsilon \|a\|$$

by Schwartz's inequality. Putting

$$u = y - \frac{\langle y,a\rangle}{\langle b,a\rangle} b,$$

we have $u \in E$ and $\langle u, a\rangle = 0$, whence $u \in H_1 \wedge E$. Moreover,

$$\|x-u\| \leqq \|x-y\| + \|y-u\| < \varepsilon + \left|\frac{\langle y,a\rangle}{\langle b,a\rangle}\right| \cdot \|b\|$$

$$\leqq \varepsilon\left(1 + \frac{\|a\| \cdot \|b\|}{\langle b,a\rangle}\right).$$

Therefore $H_1 \wedge E$ is dense in H_1.

(II) Next, we shall show that if $\langle a,b\rangle = 0$ in H then there exist two sequences $\{a_n\}$ and $\{b_n\}$ in E such that

(1) $\quad \lim_n a_n = a, \quad \lim_n b_n = b \quad$ and $\langle a_m, b_n\rangle = 0 \quad$ for all m, n.

Since $\{b\}^\perp \wedge E$ is dense in $\{b\}^\perp$ by (I) and since $a \in \{b\}^\perp$, there exists $a_1 \in \{b\}^\perp \wedge E$ such that $\|a_1 - a\| < 1$. Since $\{a, a_1\}^\perp \wedge E$ is dense in $\{a, a_1\}^\perp$ and since $b \in \{a, a_1\}^\perp$, there exists $b_1 \in \{a, a_1\}^\perp \wedge E$ such that $\|b_1 - b\| < 1$. Since $\{b, b_1\}^\perp \wedge E$ is dense in $\{b, b_1\}^\perp$ and $a \in \{b, b_1\}^\perp$, there exists $a_2 \in \{b, b_1\}^\perp \wedge E$ such that $\|a_2 - a\| < \frac{1}{2}$. Continuing this process, we obtain two sequences $\{a_n\}$ and $\{b_n\}$ having the following properties:

$$a_n \in \{b, b_1, \ldots, b_{n-1}\}^\perp \wedge E, \qquad \|a_n - a\| < \frac{1}{n},$$

$$b_n \in \{a, a_1, \ldots, a_{n-1}, a_n\}^\perp \wedge E, \qquad \|b_n - b\| < \frac{1}{n}.$$

Then $\{a_n\}$ and $\{b_n\}$ satisfy (1).

(III) Let a be a non-zero element of H. We shall show that $a \in E$. Since E is dense in H, there exists $c \in E$ such that $\langle c, a\rangle \neq 0$. Putting

(2) $$b = a - \frac{\langle a,a\rangle}{\langle c,a\rangle} c,$$

we have $\langle b, a\rangle = 0$. By (II) there exist two sequences $\{a_n\}$ and $\{b_n\}$ in E satisfying (1). Put

$$A = \{x \in E; \langle x, b_n\rangle = 0 \text{ for every } n\}.$$

Then A is an E-closed subspace of E and it is a subspace of H. Hence the closure $\bar{A}$ in H is a closed subspace of H. Let P be the projection operator on $\bar{A}$. Since $a_m \in A$ for every m, we have $a \in \bar{A}$, whence $Pa = a$. Since $b_n \in A^\perp = (\bar{A})^\perp$ for every n, we have $b \in (\bar{A})^\perp$, whence $Pb = 0$. Therefore, from (2) we have

(3) $$a = \frac{\langle a,a\rangle}{\langle c,a\rangle} Pc.$$

On the other hand, since $A \in L_E(E)$ and since $L_E(E)$ is orthomodular, we have $(A, A^0)M^*$ in $L_E(E)$, and then by (33.4.1) we have

$$A + A^0 = A \vee A^0 = E.$$

Hence there exist $c_1, c_2 \in E$ such that

$$c = c_1 + c_2 \quad \text{with } c_1 \in A \quad \text{and } c_2 \in A^0.$$

Since $A^0 = A^{\perp} \wedge E$, we have $c_2 \in A^{\perp} = (\bar{A})^{\perp}$, whence $Pc_2 = 0$. Hence

$$Pc = Pc_1 = c_1 \in E.$$

From (3) we have $a \in E$. Consequently, we have $E = H$, whence E is a Hilbert space. □

Theorem (34.10). *Let E be an inner product space. The following statements are equivalent.*

(α) *E is a Hilbert space.*

(β) *In $L_c(E)$, $(A, B)M$ implies $(A, B)M^*$* ($L_c(E)$ is cross-symmetric).

(γ) *In $L_c(E)$, $(A, B)M$ and $(A, B)M^*$ are equivalent* ($L_c(E)$ is cross-symmetric and dual cross-symmetric).

Proof. The implication (α)$\Rightarrow$(γ) follows from (34.8), and (γ)$\Rightarrow$(β) is trivial.

(β)$\Rightarrow$(α). Let H be the completion of E, and assume that $H \neq E$. It suffices to show that there exists a modular pair (A, B) in $L_c(E)$ which is not dual-modular. Let a be an element of H with $a \notin E$. As in (III) of the previous proof, we can take an element $b \in H$ such that

(1) $$a - b \in E \quad \text{and} \quad \langle a, b \rangle = 0,$$

and we can take two sequences $\{a_n\}$ and $\{b_n\}$ in E such that

(2) $$a_n \to a, \quad b_n \to b \quad \text{and} \quad \langle a_m, b_n \rangle = 0 \quad \text{for all } m, n.$$

Denote by A and B the closed subspaces of E generated by $\{a_n\}$ and $\{b_n\}$ respectively.

For any $x \in A$ and $y \in B$, since $\langle x, y \rangle = 0$ by (2), we have

$$\|x + y\|^2 = \|x\|^2 + \|y\|^2.$$

Hence the mapping $(x, y) \to x + y$ is a topological isomorphism of $A \times B$ onto $A + B$. Therefore, the pair (A, B) is modular by (32.11).

By (2) we have $a \in \bar{A}$ (the closure of A in H) and $b \in \bar{B}$. Moreover we have $\langle x, y \rangle = 0$ for all $x \in \bar{A}$ and $y \in \bar{B}$, whence $\bar{A} \cap \bar{B} = \{0\}$. If $A + B$ were closed in E, then since $a_n - b_n \to a - b \in E$ by (1) and (2), we would have $a - b \in A + B$, whence

$$a - b = a_1 + b_1 \quad \text{for some } a_1 \in A \quad \text{and } b_1 \in B.$$

Since $a-a_1=b+b_1\in\bar{A}\cap\bar{B}=\{0\}$, we would have $a=a_1\in A\subset E$, a contradiction. Therefore $A+B$ is not closed in E and hence (A,B) is not dual-modular in $L_c(E)$. ▯

PROBLEM 8. (Mackey) Let E be a normed space. If in $L_c(E)$ $(A,B)M$ and $(A,B)M^*$ are equivalent, then is E necessarily complete?

References for Chapter VII

For Section 31: H. H. Schaefer [1], S. Maeda [6] and [7].

For Section 32: G. W. Mackey [1], S. Maeda [7], H. H. Schaefer [1], A. Martineau [1].

For Section 33: S. Maeda [6], J. E. McLaughlin [1], R. Baer [1].

For Section 34: G. Birkhoff and J. von Neumann [1], M. D. MacLaren [1], R. Baer [1], P. R. Halmos [1], C. Piron [1]; (34.9) is due to I. Amemiya and H. Araki [1], (34.10) is due to S. S. Holland, Jr. [3].

Chapter VIII

Orthomodular Symmetric Lattices

35. Relatively Complemented Symmetric Lattices with Duality

Lemma (35.1). *Let J be an ideal of a lattice L, and assume that every element of J is modular. If $x,y\in J$ and $x\leqq a\vee y$ in L, then there exists an element $u\in J$ such that $x\leqq u\vee y$ and $u\leqq a$.*

Proof. Let $x,y\in J$ and $x\leqq a\vee y$. We put $u=a\wedge(x\vee y)$. Then we have $u\in J$ and $u\leqq a$. Since $(a,x\vee y)M$ by the assumption, we have

$$u\vee y=y\vee\{a\wedge(x\vee y)\}=(y\vee a)\wedge(x\vee y)\geqq x. \quad \square$$

Definition (35.2). A subset S of a lattice L is called *join-dense* in L when

$$a=\bigvee(x\in S;\, x\leqq a) \quad \text{for every } a\in L.$$

For instance, if L is atomistic then a set including all atoms of L is join-dense.

Definition (35.3). Let L be a lattice with 0. An ideal J of L is called a *p-ideal* when any element of L which is subperspective to an element in J belongs to J.

When L is relatively complemented, it follows from (6.4) that an ideal J is a p-ideal if $a\sim b$ and $b\in J$ imply $a\in J$.

Theorem (35.4). *Let L be a relatively complemented lattice with 0, and assume that*

(J) *L has a join-dense p-ideal J whose elements are all modular.*

Then $(a,b)M^*$ in L if and only if the following statement holds:

(35.4.1) *If $x\in J$ and $x\leqq a\vee b$, then there exist elements $u,v\in J$ such that $x\leqq u\vee v$, $u\leqq a$ and $v\leqq b$.*

Proof. (I) Assume that (35.4.1) holds. Let $c\geqq b$. We shall prove that

(1) $\quad$ if $x\in J$ and $x\leqq c\wedge(a\vee b)$ then $x\leqq(c\wedge a)\vee b$.

By (35.4.1) there exist $u, v \in J$ such that $x \leqq u \vee v$, $u \leqq a$, and $v \leqq b$. Since $v \vee x \in J$ is a modular element, we have

$$v \vee \{u \wedge (v \vee x)\} = (v \vee u) \wedge (v \vee x) \geqq x.$$

Moreover, $v \vee x \leqq b \vee c = c$. Hence we have

$$x \leqq v \vee (u \wedge c) \leqq b \vee (a \wedge c).$$

Thus (1) holds. Since J is join-dense, it holds that

$$c \wedge (a \vee b) \leqq (c \wedge a) \vee b.$$

Therefore $(a, b)M^*$ holds.

(II) Assume $(a, b)M^*$ and let $x \in J$ with $x \leqq a \vee b$. We put $c = b \vee x$. Since $(a, b)M^*$ and $c \leqq a \vee b$, we have

$$(c \wedge a) \vee b = c \wedge (a \vee b) = c = b \vee x.$$

We take a complement u of $a \wedge b$ in $L[0, c \wedge a]$. Then

$$\begin{aligned} u \wedge b &= u \wedge (c \wedge a) \wedge b = 0 \quad \text{and} \\ u \vee b &= u \vee (a \wedge b) \vee b = (c \wedge a) \vee b = b \vee x. \end{aligned} \tag{2}$$

Hence u is subperspective to x. Since J is a p-ideal, we have $u \in J$. By (2) we have $x \leqq u \vee b$. Hence it follows from (35.1) that there exists $v \in J$ such that $x \leqq u \vee v$ and $v \leqq b$. Thus (35.4.1) holds. □

Corollary (35.5). *Let L be a relatively complemented lattice with* 0 *satisfying the condition* (*J*) *in* (35.4).

(35.5.1) *L is M^*-symmetric.*

(35.5.2) *If $a \in J$ then $(a, b)M$, $(b, a)M$, $(a, b)M^*$ and $(b, a)M^*$ for every $b \in L$.*

Proof. (35.5.1) follows from the fact that the statement (35.4.1) is symmetric for a and b.

If $a \in J$ then $(b, a)M$ holds for every $b \in L$ since a is modular. It follows from (35.1) and (35.4) that $(a, b)M^*$ and $(b, a)M^*$ hold. Moreover, $(a, b)M$ holds by (1.2). □

Theorem (35.6). *If a relatively complemented lattice L with* 0 *and* 1 *satisfies the condition* (*J*), *then the set*

$$J^* = \{x \in L; x \text{ has a complement } x' \in J\}$$

is a join-dense p-ideal of the dual L^* and every element $x \in J^*$ is modular in L^*.

Proof. (I) If $x \in J^*$ then every complement of x belongs to J, since J is a p-ideal. Moreover,

(1) $\qquad$ if $x \vee u = 1$ for some $u \in J$ then $x \in J^*$,

since a complement of x is subperspective to u; and

(2) $\qquad$ if $u \wedge x = 0$ for some $x \in J^*$ then $u \in J$,

since u is subperspective to a complement of x.

We shall show that J^* is an ideal of L^*. It is evident by (1) that if $a \in J^*$ and $a \leqq b$, then $b \in J^*$. Let $a, b \in J^*$. Then there exists $u \in J$ such that $a \vee u = 1$. Taking a complement w of $a \wedge b$ in $L[0, a]$, we have $w \wedge b = w \wedge a \wedge b = 0$, whence $w \in J$ by (2). Since $w \vee u \in J$ and $(a \wedge b) \vee w \vee u = a \vee u = 1$, we have $a \wedge b \in J^*$ by (1). Hence J^* is an ideal of L^*.

(II) Let $a \in J^*$ and let b be subperspective to a in L^*. Then $a \vee u = 1$ for some $u \in J$, and $b \geqq a \wedge x$ and $b \vee x = 1$ for some $x \in L$. Taking a complement y of $a \wedge x$ in $L[0, x]$, we have

$$a \wedge y = a \wedge x \wedge y = 0 \quad \text{and} \quad b \vee y = b \vee (a \wedge x) \vee y = b \vee x = 1.$$

Hence $y \in J$ by (2), and hence $b \in J^*$ by (1). Therefore J^* is a p-ideal of L^*.

(III) We shall show that if $a < 1$, then there exists $b \in J^*$ such that $a \leqq b < 1$. Let a' be a complement of a. Since $a' > 0$ and since J is join-dense, there exists $u \in J$ such that $0 < u \leqq a'$. Taking a complement b of $a \vee u$ in $L[a, 1]$, we have $b < 1$, since otherwise $a \vee u = a$, whence $u \leqq a \wedge a' = 0$, a contradiction. Moreover, $b \in J^*$ since $b \vee u = b \vee a \vee u = 1$.

(IV) Let a be an arbitrary element of L. Evidently a is a lower bound of the set $S = \{x \in J^*; x \geqq a\}$. Let b be a lower bound of S. If we had $a < a \vee b$, then taking a complement c of $a \vee b$ in $L[a, 1]$, by (III) we would have an element $x \in J^*$ such that $c \leqq x < 1$. Then $x \geqq b$, since $x \in S$. Hence $x \geqq b \vee c = a \vee b \vee c = 1$, a contradiction. Thus we have $a = a \vee b \geqq b$ for every lower bound b of S. Therefore $a = \bigwedge(x \in J^*; x \geqq a)$, which means that J^* is join-dense in L^*.

(V) Let $a \in J^*$. We shall show $(b, a)M^*$ for every $b \in L$. Let $c \geqq a$, and let u be a complement of $c \wedge b$ in $L[0, b]$. Since $u \wedge a \leqq u \wedge c = u \wedge b \wedge c = 0$, we have $u \in J$ by (2). Since $(u, (c \wedge b) \vee a)M^*$ by (35.5.2), we have

$$c \wedge (b \vee a) = c \wedge \{u \vee (c \wedge b) \vee a\} = (c \wedge u) \vee \{(c \wedge b) \vee a\} = (c \wedge b) \vee a.$$

Hence $(b, a)M^*$ holds for every b, whence a is modular in L^*. $\square$

Corollary (35.7). *Let L be a relatively complemented lattice with* 0 *and* 1 *satisfying* (*J*).

(35.7.1) *L is M-symmetric and M^*-symmetric.*

(35.7.2) *If $a \in J \cup J^*$ then $(a, b)M$, $(b, a)M$, $(a, b)M^*$ and $(b, a)M^*$ for every $b \in L$.*

Proof. It follows from (35.6) that L^* satisfies (J). Hence (35.5.1) and (35.5.2) hold both in L and in L^*. ▯

Remark (35.8). If a lattice L with 0 satisfies (J), then L must be finite-modular; because, since the ideal J is join-dense, it contains all atoms and hence it contains all finite elements.

Moreover, an AC-lattice L satisfies (J) if and only if L is finite-modular; because, if L is finite-modular, then the join-dense set $\mathscr{F}(L)$ of finite elements of L forms a p-ideal by (8.8) and (11.9), and every element of $\mathscr{F}(L)$ is modular.

Lemma (35.9). *Let J be a join-dense set in a lattice L. If $(a, b \vee x)M$ for every $x \in J$ with $x \nleqq b$ then $(a,b)M^*$.*

Proof. Let $c \geqq b$ and let $x \in J$ with $x \leqq c \wedge (a \vee b)$. We shall show that $x \leqq (c \wedge a) \vee b$. This is evident when $x \leqq b$. When $x \nleqq b$, by $(a, b \vee x)M$ we have

$$x \leqq (b \vee a) \wedge (b \vee x) = b \vee \{a \wedge (b \vee x)\} \leqq b \vee (a \wedge c) = (c \wedge a) \vee b.$$

Since J is join-dense, we have $c \wedge (a \vee b) \leqq (c \wedge a) \vee b$. Hence $(a,b)M^*$ holds. ▯

Lemma (35.10). *Let L be a relatively complemented lattice with 0 satisfying the condition (J). In L, if $(a,b)M^*$ then $(a \vee x, b \vee y)M^*$ for all $x, y \in J$.*

Proof. This can be proved by the same way as (27.12). ▯

Theorem (35.11). *Let a and b be elements of a relatively complemented lattice L with 0 and 1 satisfying the condition (J). The following three statements are equivalent.*

(α) $(a,b)M$ *and* $(a,b)M^*$.
(β) $(a, b \vee x)M$ *for every* $x \in J$ *with* $x \nleqq b$.
(γ) $(a, b \wedge y)M^*$ *for every* $y \in J^*$ *with* $y \ngeqq b$.

Proof. (γ)⇒(α). It follows from (γ) that in L^* $(a, b \vee y)M$ for every $y \in J^*$ with $y \nleqq b$. Since J^* is join-dense in L^* by (35.6), it follows from (35.9) that $(a,b)M^*$ in L^*. Hence $(a,b)M$ in L.

To prove $(a,b)M^*$, we may assume $b \neq 0$. Since J^* is join-dense in L^*, we have $\bigwedge(y; y \in J^*) = 0$ in L. Hence there exists $y \in J^*$ such that $y \ngeqq b$. Taking a complement x of $b \wedge y$ in $L[0,b]$, we have $x \wedge y = x \wedge b \wedge y = 0$, whence $x \in J$. Since $(a, b \wedge y)M^*$ by (γ) and since $b = (b \wedge y) \vee x$, we have $(a,b)M^*$ by (35.10).

(α)⇒(γ). Let $y \in J^*$ with $y \ngeqq b$, and let $x \in J$ with $x \leqq a \vee (b \wedge y)$. Since $(a,b)M^*$ by (α) and since $x \leqq a \vee b$, there exist $u_1, v_1 \in J$ such that $x \leqq u_1 \vee v_1, u_1 \leqq a$ and $v_1 \leqq b$. Putting $v_2 = (u_1 \vee x) \wedge v_1$, we have $v_2 \in J$ and

$$(1) \qquad \begin{aligned} u_1 \vee v_2 &= u_1 \vee \{v_1 \wedge (u_1 \vee x)\} \\ &= (u_1 \vee v_1) \wedge (u_1 \vee x) \geqq x. \end{aligned}$$

We take a complement w of $a \wedge b \wedge y$ in $L[0, a \wedge b]$. Then since $w \wedge y = w \wedge (a \wedge b) \wedge y = 0$, we have $w \in J$. Since $(a,b)M$ by (α), we have

$$(b \wedge y) \vee w = (b \wedge y) \vee (a \wedge b \wedge y) \vee w = (b \wedge y) \vee (a \wedge b) = \{(b \wedge y) \vee a\} \wedge b.$$

Hence we have

$$v_2 = (u_1 \vee x) \wedge v_1 \leqq \{a \vee (b \wedge y)\} \wedge b = (b \wedge y) \vee w.$$

By (35.1) there exists $v \in J$ such that $v_2 \leqq v \vee w$ and $v \leqq b \wedge y$. By (1) we have

$$x \leqq u_1 \vee v_2 \leqq u_1 \vee v \vee w, \quad \text{where } u_1 \vee w \leqq a \quad \text{and } v \leqq b \wedge y.$$

Therefore $(a, b \wedge y)M^*$ holds.

(α)⇔(β). (β) holds in L if and only if (γ) holds in L^*, since $J^{**} = J$. Evidently (α) holds in L if and only if it holds in L^*. Therefore the equivalence of (α) and (γ) in L^* implies the equivalence of (α) and (β) in L. □

Corollary (35.12). *Let L be a relatively complemented lattice with 0 and 1 satisfying (J).*

(35.12.1) *In L the following statements are equivalent.*
(α) $(a,b)M$ *implies* $(a,b)M^*$ (L is cross-symmetric).
(β) *If* $(a,b)M$ *then* $(a, b \vee x)M$ *for every* $x \in J$.

(35.12.2) *In L the following statements are equivalent.*
(α*) $(a,b)M^*$ *implies* $(a,b)M$ (L is dual cross-symmetric).
(β*) *If* $(a,b)M^*$ *then* $(a, b \wedge y)M^*$ *for every* $y \in J^*$.

Proof. If $(a,b)M$, then by the equivalence of (α) and (β) in (35.11), $(a,b)M^*$ holds if and only if $(a, b \vee x)M$ holds for every $x \in J$. Hence (α) and (β) are equivalent. Similarly (α*) and (β*) are equivalent. □

Remark (35.13). Let L be an orthomodular lattice satisfying the condition (J). Since L is relatively complemented, the results (35.4), (35.5) and (35.10) are available for L. Moreover, it is easy to show that $J^* = \{x^\perp; x \in J\}$, and then the results (35.6) and (35.7) are evident since L is self-dual. The results (35.11) and (35.12) are also available for L.

Corollary (35.14). *Let L be an orthomodular lattice satisfying (J). L is O-symmetric if and only if, in L, $(a,b)M$ implies $(a, b \vee x)M$ for every $x \in J$.*

Proof. This is a direct consequence of (35.12.1), since the condition (α) in (35.12) is equivalent to O-symmetry of L. □

Remark (35.15). A complete relatively semi-orthocomplemented lattice L, whose semi-orthogonality is ortho-continuous, is called a *dimension lattice* when there exists an equivalence relation "$\sim$" in L satisfying the following conditions:

(ED 1) $a \sim 0$ implies $a = 0$.
(ED 2) If $a \sim b_1 \oplus b_2$ (we denote by $\oplus$ the join of a semi-orthogonal family), then there exists a decomposition $a = a_1 \oplus a_2$ with $a_i \sim b_i$ $(i = 1, 2)$.
(ED 3) If a and b are perspective, then $a \sim b$.
(ED 4) If $a = \oplus_\alpha a_\alpha$, $b = \oplus_\alpha b_\alpha$, $a_\alpha \sim b_\alpha$ for every α and $a \perp b$, then $a \sim b$.
(ED 5) If $a = a_1 \oplus a_2$, $b = b_1 \oplus b_2$ and $a_i \sim b_i$ $(i = 1, 2)$, then $a \sim b$.

(The condition (ED 5) may be omitted. See S. Maeda [2], Introduction.)

An element a of a dimension lattice L is called *finite* when $L[0, a]$ has no infinite semi-orthogonal family $\{a_\alpha; \alpha \in I\}$ such that $a_\alpha \sim a_\beta$ for all $\alpha, \beta \in I$ and $a_\alpha \neq 0$. It can be proved that a is finite if and only if $a \sim a_1 \leqq a$ imply $a_1 = a$ and that the set J of all finite elements of L forms a p-ideal (see S. Maeda [2], Lemma 4.6 and Theorem 6.1). Moreover, it is easy to show by (ED 3) that every element of J is modular.

It can be proved (see S. Maeda [2], § 2 and § 3) that there exists a central element z of L having the following two properties:

(1) Every element in $L[0, z]$ is the join of a family of finite elements of L, and

(2) $L[0, z']$ has no non-zero finite element, where z' is the complement of z.

We say that L is *locally finite* when $z = 1$. By the above considerations, we have the following conclusion:

If a dimension lattice L is locally finite, then L satisfies the condition (J) and hence L is M-symmetric and M*-symmetric by (35.7).

Remark (35.16). Let L be a complete orthomodular lattice. It is evident that L is relatively semi-orthocomplemented when a semi-orthogonality in L is defined by the orthogonality, and that this semi-orthogonality is ortho-continuous. It can be proved that the following four statements are equivalent.

(α) L satisfies the condition (J).
(β) L is M-symmetric and the set

$$\{a \in L; a \text{ is modular and } L[0, a] \text{ is modular}\}$$

is join-dense in L (L is nearly modular in the sense of MacLaren [2]).

(γ) L has a join-dense modular ideal (L is locally modular in the sense of Ramsay [1]).

(δ) L is a locally finite dimension lattice.

The implication $(\alpha)\Rightarrow(\beta)$ follows from (35.7). $(\beta)\Rightarrow(\delta)$ is proved in MacLaren [2]. $(\delta)\Rightarrow(\alpha)$ follows from (35.15). $(\alpha)\Rightarrow(\gamma)$ is trivial and $(\gamma)\Rightarrow(\alpha)$ is proved in Ramsay [1].

Of course, the proofs of $(\beta)\Rightarrow(\delta)$ and $(\delta)\Rightarrow(\alpha)$ belong to the dimension theory of lattices. The proof of $(\gamma)\Rightarrow(\alpha)$ also is based on some results in the dimension theory.

Remark (35.17). A dimension lattice L is called of type III when L has no non-zero finite element (or equivalently, $z=0$ in (35.15)).

It can be proved that a dimensional lattice of type III is not necessarily M-symmetric. In fact, we can construct a complete orthomodular lattice L which has no atoms and is not M-symmetric, from two irreducible complete orthocomplemented modular lattices with no atoms, by pasting in the sense of Greechie [1]. If we define in L the following equivalence relation:

$$a\sim b \quad \text{when either } a=b=0 \quad \text{or} \quad a>0 \text{ and } b>0,$$

then it is easy to see that L becomes a dimension lattice of type III. This lattice is not M-symmetric.

EXERCISE 35.1. Let L be a relatively complemented lattice with 0 satisfying (J). Prove that in L if $(a,b)M$ and $(a,b)M^*$ then $(a,x)M^*$ for any $a\wedge b\leqq x\leqq b$.

36. Commutativity in Orthomodular Lattices

Theorem (36.1). *An orthocomplemented lattice L is orthomodular if and only if L satisfies the following condition:*

(36.1.1) *If* $a=(a\wedge b)\vee(a\wedge b^{\perp})$ *then* $b=(b\wedge a)\vee(b\wedge a^{\perp})$.

Proof. (I) Let L be orthomodular. If $a=(a\wedge b)\vee(a\wedge b^{\perp})$, then $a\vee b^{\perp}=(a\wedge b)\vee b^{\perp}$. Since $(a\wedge b,(a\wedge b)^{\perp})M$ we have

$$(a\vee b^{\perp})\wedge(a\wedge b)^{\perp}=\{b^{\perp}\vee(a\wedge b)\}\wedge(a\wedge b)^{\perp}=b^{\perp}.$$

Taking orthocomplements, $b=(a^{\perp}\wedge b)\vee(a\wedge b)$. Therefore (36.1.1) holds.

(II) Let (36.1.1) hold. If $a\leqq b$, then since $a\wedge b^{\perp}\leqq a\wedge a^{\perp}=0$, we have

$$(a\wedge b)\vee(a\wedge b^{\perp})=a\wedge b=a.$$

By (36.1.1) we have

$$b=(b\wedge a)\vee(b\wedge a^{\perp})=a\vee(b\wedge a^{\perp}).$$

Hence L is orthomodular by (29.13). □

Definition (36.2). Let a and b be elements of an orthomodular lattice L. We say that a *commutes* with b and we write $a\,C\,b$ when

$$a=(a\wedge b)\vee(a\wedge b^{\perp}).$$

By (36.1), $a\,C\,b$ if and only if $b\,C\,a$.

Lemma (36.3). *Let a and b be elements of an orthomodular lattice.*

(36.3.1) $$aCb \Leftrightarrow bCa \Leftrightarrow bCa^{\perp} \Leftrightarrow a^{\perp}Cb \Leftrightarrow a^{\perp}Cb^{\perp} \Leftrightarrow b^{\perp}Ca^{\perp} \Leftrightarrow b^{\perp}Ca \Leftrightarrow aCb^{\perp}.$$

(36.3.2) $a\leqq b$ *implies* aCb.

(36.3.3) aCb *implies* $(a,b)M$ *and* $(a,b)M^*$.

(36.3.4) $a\perp b$ *if and only if* $a\wedge b=0$ *and* aCb.

Proof. (36.3.1) and (36.3.2) are evident. Assume aCb and let $c\leqq b$. Since $(a\wedge b^{\perp},b)M$ by the orthomodularity and since $c\vee(a\wedge b)\leqq b$, we have

$$\begin{aligned}(c\vee a)\wedge b&=\{c\vee(a\wedge b)\vee(a\wedge b^{\perp})\}\wedge b=\{c\vee(a\wedge b)\}\vee\{(a\wedge b^{\perp})\wedge b\}\\&=c\vee(a\wedge b).\end{aligned}$$

Hence $(a,b)M$ holds. Moreover, aCb implies $a^{\perp}Cb^{\perp}$ by (36.3.1) and hence implies $(a^{\perp},b^{\perp})M$. Hence $(a,b)M^*$ holds.

Now we shall prove (36.3.4). If $a\perp b$, then $a\wedge b=0$ and moreover $a\wedge b^{\perp}=a$. Hence aCb. Conversely, if $a\wedge b=0$ and aCb, then

$$a=(a\wedge b)\vee(a\wedge b^{\perp})=a\wedge b^{\perp}\leqq b^{\perp},$$

whence $a\perp b$. □

Lemma (36.4). *In an orthomodular lattice L, if* a_1Cb *and* a_2Cb *then* $a_1\vee a_2Cb$ *and* $a_1\wedge a_2Cb$. *More generally, if* $a_\alpha Cb$ *for every* $\alpha\in I$ *then*

$$\bigvee(a_\alpha;\alpha\in I)Cb \quad \textit{and} \quad \bigwedge(a_\alpha;\alpha\in I)Cb,$$

providing the elements involved exist.

Proof. We shall prove the last statement. Assume $a_\alpha Cb$ for every $\alpha\in I$. When $\bigvee(a_\alpha;\alpha\in I)$ exists, for every $\beta\in I$ we have

$$a_\beta=(a_\beta\wedge b)\vee(a_\beta\wedge b^{\perp})\leqq(\bigvee_\alpha a_\alpha\wedge b)\vee(\bigvee_\alpha a_\alpha\wedge b^{\perp}).$$

Hence

$$\bigvee_\alpha a_\alpha \leqq (\bigvee_\alpha a_\alpha \wedge b) \vee (\bigvee_\alpha a_\alpha \wedge b^\perp) \leqq \bigvee_\alpha a_\alpha .$$

Thus $\bigvee_\alpha a_\alpha C b$ holds. When $\bigwedge(a_\alpha; \alpha \in I)$ exists, we have $\bigvee_\alpha a_\alpha^\perp = (\bigwedge_\alpha a_\alpha)^\perp$. Since $a_\alpha^\perp C b$ for every $\alpha \in I$, we have $(\bigwedge_\alpha a_\alpha)^\perp C b$ as above. Hence $\bigwedge_\alpha a_\alpha C b$. ▯

Definition (36.5). In an orthomodular lattice, if $a \perp b$ then we write $a \oplus b$ for the element $a \vee b$, and if $a \leqq b$ then we write $b - a$ for the element $b \wedge a^\perp$ (which is the relative orthocomplement of a in $L[0, b]$).

Lemma (36.6). *In an orthomodular lattice,*

(36.6.1) *if* $a \leqq b$ *and* $b - a = 0$ *then* $a = b$,

(36.6.2) $a C b$ *if and only if* $a \wedge b = a \wedge (b \vee a^\perp)$ $(= a - (a \wedge b^\perp))$.

Proof. (I) If $a \leqq b$ then by (δ) of (29.13) we have $b = a \oplus (b - a)$. Hence $b - a = 0$ implies $b = a$.

(II) If $a C b$, then

$$\{a \wedge (b \vee a^\perp)\} - (a \wedge b) = a \wedge (b \vee a^\perp) \wedge (a \wedge b)^\perp = a \wedge \{(a \wedge b^\perp) \vee (a \wedge b)\}^\perp$$
$$= a \wedge a^\perp = 0.$$

Hence $a \wedge b = a \wedge (b \vee a^\perp)$ by (I). Conversely, if $a \wedge b = a \wedge (b \vee a^\perp) = a - (a \wedge b^\perp)$, then

$$(a \wedge b) \oplus (a \wedge b^\perp) = \{a - (a \wedge b^\perp)\} \oplus (a \wedge b^\perp) = a,$$

whence $a C b$. ▯

Theorem (36.7). *Let a, b and c be elements of an orthomodular lattice L. If some one of these three elements commutes with the other two, then* $\{a, b, c\}$ *is a distributive triple.*

Proof. (I) We may assume without loss of generality that c commutes with a and b. There are six different laws we have to verify; these are $(a, b, c)D$, $(a, c, b)D$, $(b, c, a)D$, $(a, b, c)D^*$, $(a, c, b)D^*$ and $(b, c, a)D^*$.

(II) We shall prove $(a, b, c)D$. Evidently

$$(a \vee b) \wedge c \geqq (a \wedge c) \vee (b \wedge c).$$

Since $c C a^\perp$ and $c C b^\perp$ by (36.3.1), we have

$$c \wedge (a^\perp \vee c^\perp) = c \wedge a^\perp \quad \text{and} \quad c \wedge (b^\perp \vee c^\perp) = c \wedge b^\perp$$

by (36.6.2). Hence

$$\{(a \vee b) \wedge c\} - \{(a \wedge c) \vee (b \wedge c)\} = (a \vee b) \wedge c \wedge (a^\perp \vee c^\perp) \wedge (b^\perp \vee c^\perp)$$
$$= (a \vee b) \wedge c \wedge a^\perp \wedge b^\perp$$
$$= (a \vee b) \wedge (a \vee b)^\perp \wedge c = 0.$$

Therefore $(a, b, c)D$ holds by (36.6.1).

(III) We shall prove $(a, c, b)D$. Evidently

$$(a \vee c) \wedge b \geqq (a \wedge b) \vee (c \wedge b).$$

Since $bCc^\perp$ and $c^\perp Ca$, we have

$$b \wedge (c^\perp \vee b^\perp) = b \wedge c^\perp \quad \text{and} \quad c^\perp \wedge (a \vee c) = c^\perp \wedge a.$$

Hence

$$\{(a \vee c) \wedge b\} - \{(a \wedge b) \vee (c \wedge b)\} = (a \vee c) \wedge b \wedge (a^\perp \vee b^\perp) \wedge (c^\perp \vee b^\perp)$$
$$= (a \vee c) \wedge b \wedge c^\perp \wedge (a^\perp \vee b^\perp)$$
$$= a \wedge b \wedge c^\perp \wedge (a \wedge b)^\perp = 0.$$

Therefore $(a, c, b)D$ holds. Similarly we can prove $(b, c, a)D$.

(IV) Since $c^\perp$ commutes with $a^\perp$ and $b^\perp$, we can prove $(a^\perp, b^\perp, c^\perp)D$, $(a^\perp, c^\perp, b^\perp)D$ and $(b^\perp, c^\perp, a^\perp)D$ as above. Taking orthocomplements, we get $(a, b, c)D^*, (a, c, b)D^*$ and $(b, c, a)D^*$.   □

Corollary (36.8). *In an orthomodular lattice L, aCb if and only if*

(36.8.1)   *the sublattice generated by $\{a, a^\perp, b, b^\perp\}$ is distributive.*

Proof. If (36.8.1) holds, then $(b, b^\perp, a)D$ holds, that is, aCb holds. Conversely, assume aCb. By (36.3.1) and (36.3.4), any two elements in $\{a, a^\perp, b, b^\perp\}$ commute. Then, in the sublattice generated by these four elements, any two elements commute by (36.4). Hence this sublattice is distributive by (36.7).   □

Theorem (36.9). *Let z be an element of an orthomodular lattice L. The following three statements are equivalent.*

(α) *z is a central element.*
(β) *z has a unique complement, namely $z^\perp$.*
(γ) *z commutes with every $a \in L$.*

Proof. Since L is relatively complemented, (α) and (β) are equivalent by (4.20). The implication (α) ⇒ (γ) is evident. We shall prove (γ) ⇒ (α). Since $(z, z^\perp)M$ and $(z^\perp, z)M$ hold, if aCz for every $a \in L$ then z satisfies the condition (ε) in (4.13). Hence z is central.   □

Remark (36.10). Let L be an orthomodular lattice. For any subset S of L, we put

$$C(S)=\{a\in L;\, aCb \text{ for every } b\in S\}.$$

$C(S)$ is an orthocomplemented sublattice of L by (36.3.1) and (36.4), and we have $S\subset CC(S)$. When $S=CC(S)$, we call S a *C-closed sublattice* of L. By (36.9), the center $Z(L)$ of L is a C-closed sublattice and any C-closed sublattice includes $Z(L)$.

If S is a C-closed sublattice of L, then S is orthomodular, since (δ) in (29.13) holds in S also. Moreover, the center of S is equal to the intersection $S\cap C(S)$, because by (36.9) an element a of S is in the center of S if and only if $a\in C(S)$.

If L is complete, then, for any subset S, $C(S)$ is a complete sublattice of L by (36.4). Hence any C-closed sublattice of L is a complete sublattice.

Lemma (36.11). *Let a, b and c be elements of an orthomodular lattice L. If $(a,b)M$, aCc and bCc then $(a,b\wedge c)M$, $(a\wedge c,b\wedge c)M$, $(a\vee c,b)M$ and $(a\vee c,b\vee c)M$.*

Proof. Let $(a,b)M$, aCc and bCc. Since $a\wedge bCc$ by (36.4), we have $(a\wedge b,c)M$ by (36.3.3). Hence it follows from (1.5.1) that $(a,b\wedge c)M$ and $(a\wedge c,b\wedge c)M$.

We shall prove $(a\vee c,b)M$. Let $x\leqq b$. Since $(a,b,c)T$ by (36.7) and since $x\vee(c\wedge b)\leqq b$, we have

$$\text{(1)}\qquad x\vee\{(a\vee c)\wedge b\}=x\vee(c\wedge b)\vee(a\wedge b)=\{x\vee(c\wedge b)\vee a\}\wedge b$$
$$=[x\vee\{(c\vee a)\wedge(b\vee a)\}]\wedge b.$$

Since $b\vee aCc$ and $b\vee aCa$, we have $b\vee aCc\vee a$. Moreover $xCb\vee a$ since $x\leqq b\vee a$. Hence $(x,c\vee a,b\vee a)T$ by (36.7). Therefore

$$\text{(2)}\qquad x\vee\{(c\vee a)\wedge(b\vee a)\}=(x\vee c\vee a)\wedge(x\vee b\vee a)=(x\vee a\vee c)\wedge(b\vee a).$$

By (1) and (2) we have

$$x\vee\{(a\vee c)\wedge b\}=(x\vee a\vee c)\wedge b.$$

Thus $(a\vee c,b)M$ holds.

Finally we shall prove $(a\vee c,b\vee c)M$. Let $x\leqq b\vee c$. Since $(a,b,c)T$ and $(a,b)M$, we have

$$\text{(3)}\quad x\vee\{(a\vee c)\wedge(b\vee c)\}=x\vee(a\wedge b)\vee c$$
$$=\{(x\vee c)\wedge b\}\vee(a\wedge b)\vee x\vee c=(y\wedge b)\vee x\vee c,$$

where $y=\{(x\vee c)\wedge b\}\vee a$. Since $x\vee cCc$, bCc and aCc, we have yCc, and hence $(y,b,c)T$. On the other hand, since $(b,c)M^*$ by (36.3.3) and since $x \vee c \leqq b \vee c$, $y \vee c = \{(x \vee c) \wedge b\} \vee c \vee a = \{(x \vee c) \wedge (b \vee c)\} \vee a = x \vee c \vee a$. Hence we have

(4) $$(y \wedge b) \vee c = (y \vee c) \wedge (b \vee c) = (x \vee c \vee a) \wedge (b \vee c) \geqq x.$$

By (3) and (4) we have

$$x \vee \{(a \vee c) \wedge (b \vee c)\} = (y \wedge b) \vee c = (x \vee a \vee c) \wedge (b \vee c).$$

Thus $(a \vee c, b \vee c)M$ holds. □

Theorem (36.12). *An orthomodular lattice is M-symmetric if in L*

(36.12.1) $a \wedge b = 0$, $a \vee b = 1$ *and* $(a,b)M$ *together imply* $(b,a)M$.

Proof. By (29.17), it suffices to show that L is $\perp$-symmetric. Let $a \wedge b = 0$ and $(a,b)M$, and put $c = (a \vee b)^{\perp}$. Since aCc and bCc by (36.3.4), we have $(a \vee c, b)M$ by (36.11). Moreover, since $(c, a \vee b)M$ and $c \wedge (a \vee b) = 0$, we have

$$(a \vee c) \wedge b = (a \vee c) \wedge (a \vee b) \wedge b = [a \vee \{c \wedge (a \vee b)\}] \wedge b = a \wedge b = 0,$$

and evidently we have $a \vee c \vee b = 1$. Hence, $(b, a \vee c)M$ by (36.12.1), and then $(b,a)M$ by (1.5.3). Therefore L is $\perp$-symmetric. □

Theorem (36.13). *An orthomodular lattice is O-symmetric if in L*

(36.13.1) $a \wedge b = 0$, $a \vee b = 1$ *and* $(a,b)M$ *together imply* $(b^{\perp}, a^{\perp})M$.

Proof. (I) We shall show that in L

(1) $a \wedge b = 0$ and $(a,b)M$ together imply $(b^{\perp}, a^{\perp})M$.

As in the above proof, putting $c = (a \vee b)^{\perp}$, we have $(a \vee c, b)M$, $(a \vee c) \wedge b = 0$ and $(a \vee c) \vee b = 1$. Hence, $(b^{\perp}, (a \vee c)^{\perp})M$ by (36.13.1). Since $b^{\perp}Cc$ and $(a \vee c)^{\perp}Cc$, we have $(b^{\perp} \vee c, (a \vee c)^{\perp} \vee c)M$ by (36.11). Now, since $c \leqq b^{\perp}$ and $c \leqq a^{\perp}$, we have

$$b^{\perp} \vee c = b^{\perp} \quad \text{and} \quad (a \vee c)^{\perp} \vee c = (a^{\perp} \wedge c^{\perp}) \vee c = a^{\perp} \wedge (c^{\perp} \vee c) = a^{\perp}.$$

Thus (1) has been proved.

(II) Let $(a,b)M$ and put $d = (a \wedge b)^{\perp}$. By the same way as in the proof of (1.14), we can prove that $(a,b)M$ is equivalent to $(a \wedge d, b \wedge d)M$ and that $(b,a)M^*$ is equivalent to $(b \wedge d, a \wedge d)M^*$. Since $(a \wedge d) \wedge (b \wedge d) = a \wedge b \wedge (a \wedge b)^{\perp} = 0$, it follows from (1) that $(a \wedge d, b \wedge d)M$ implies $(b \wedge d, a \wedge d)M^*$. Hence $(a,b)M$ implies $(b,a)M^*$, and hence L is O-symmetric. □

Corollary (36.14). *An orthomodular lattice L is O-symmetric if L satisfies the following condition:*

(36.14.1) *If* $a \wedge b = 0$ *and* $a \vee b = 1$ *then there exists an automorphism* θ *of L such that* $\theta(a) = b^{\perp}$ *and* $\theta(b) = a^{\perp}$.

Proof. If L satisfies (36.14.1) then it satisfies (36.13.1), since $(a,b)M$ implies $(\theta(a),\theta(b))M$. □

In the next section we shall have some examples of orthomodular lattices satisfying (36.14.1).

EXERCISE 36.1. Prove that an orthocomplemented lattice L is orthomodular if and only if L satisfies the following condition: If $a \leqq b$ then the sublattice generated by $\{a, a^{\perp}, b, b^{\perp}\}$ is distributive.

37. Lattices of Projections of Baer *-semigroups

Definition (37.1). A **-semigroup* $\mathscr{G}$ is a semigroup with an involutorial anti-automorphism $x \to x^*$ ($(xy)^* = y^* x^*$ and $x^{**} = x$). An element $e \in \mathscr{G}$ is called a *projection* when $e = e^2 = e^*$. The set of all projections of $\mathscr{G}$ is denoted by $P(\mathscr{G})$.

We define an order in $P(\mathscr{G})$ by $e \leqq f$ if and only if $ef = e$ (or equivalently $fe = e$). It is easy to see that $P(\mathscr{G})$ is a partially ordered set. If $\mathscr{G}$ has a two-sided zero element 0 then 0 is the least element of $P(\mathscr{G})$, and if $\mathscr{G}$ has a two-sided unit element 1 then 1 is the greatest element of $P(\mathscr{G})$.

Lemma (37.2). *Let e and f be projections in a *-semigroup $\mathscr{G}$. Then $e\mathscr{G} \subset f\mathscr{G}$ if and only if $e \leqq f$.*

Proof. If $e \leqq f$, then since $e = fe$, we have $e\mathscr{G} \subset f\mathscr{G}$. Conversely, if $e\mathscr{G} \subset f\mathscr{G}$, then since $e = eee \in e\mathscr{G} \subset f\mathscr{G}$, there exists $x \in \mathscr{G}$ such that $e = fx$, Hence

$$fe = ffx = fx = e. \quad \square$$

Definition (37.3). A *-semigroup $\mathscr{G}$ with 0 is called a *Baer *-semigroup* when for each element $a \in \mathscr{G}$ there exists a projection e of $\mathscr{G}$ such that

$$\{x \in \mathscr{G}; ax = 0\} = e\mathscr{G}.$$

It follows from (37.2) that e is uniquely determined by a. We denote e by a'.

Lemma (37.4). *Any Baer *-semigroup $\mathscr{G}$ has a two-sided unit element* 1 *and it holds that* $0' = 1$ *and* $1' = 0$.

Proof. Since $0'\mathscr{G} = \{x \in \mathscr{G}; 0x = 0\} = \mathscr{G}$, we have $0'x = x$ for every $x \in \mathscr{G}$, and moreover $x0' = (0'x^*)^* = x^{**} = x$. Hence $0'$ is the unit element of $\mathscr{G}$. We have $1' = 0$ since $1'\mathscr{G} = \{x \in \mathscr{G}; 1x = 0\} = \{0\}$. □

Lemma (37.5). *Let a and b be elements of a Baer *-semigroup $\mathscr{G}$ and let e and f be projections of $\mathscr{G}$.*

(37.5.1) $aa'=0$ *and* $a'a^*=0$.

(37.5.2) $ae=0$ *implies* $e\leqq a'$.

(37.5.3) $a'\leqq(ba)'$.

(37.5.4) $e\leqq f$ *implies* $f'\leqq e'$.

(37.5.5) $a=aa''$ *and* $e\leqq e''$.

(37.5.6) $a'=a'''$.

(37.5.7) $ab=0$ *if and only if* $a''b=0$.

(37.5.8) $ea=ae$ *implies* $e'a=ae'$.

Proof. (I) We have $aa'=0$ since $a'\in a'\mathscr{G}=\{x\in\mathscr{G}; ax=0\}$. Hence $a'a^*=(aa')^*=0$.

(II) If $ae=0$, then $e\in a'\mathscr{G}$, whence $a'e=e$. Hence $e\leqq a'$.

(III) Since $baa'=0$ by (I), we have $a'\leqq(ba)'$ by (II).

(IV) If $e\leqq f$, then by (III) we have $f'\leqq(ef)'=e'$.

(V) Since $a'a^*=0$ by (I), we have $a^*\in a''\mathscr{G}$. Hence $aa''=(a''a^*)^*=a^{**}=a$. Especially $ee''=e$, whence $e\leqq e''$.

(VI) We have $a'\leqq a'''$ by (V). On the other hand, since $a''x=0$ implies $ax=aa''x=0$ by (V), we have $a'''\mathscr{G}\subset a'\mathscr{G}$. Hence $a'''\leqq a'$ by (37.2).

(VII) By (VI) we have $\{x\in\mathscr{G}; ax=0\}=a'\mathscr{G}=a'''\mathscr{G}=\{x\in\mathscr{G}; a''x=0\}$.

(VIII) If $ea=ae$, then since $eae'=aee'=0$ by (I), we have $ae'\in e'\mathscr{G}$. Hence $e'ae'=ae'$. Since $ea^*=(ae)^*=(ea)^*=a^*e$, similarly we have $e'a^*e'=a^*e'$, whence $e'ae'=e'a$. Therefore $ae'=e'a$. ▯

Definition (37.6). Let $\mathscr{G}$ be a Baer *-semigroup. A projection e of $\mathscr{G}$ is said to be *closed* when $e=e''$. By (37.5.6) a projection e is closed if and only if $e=a'$ for some $a\in\mathscr{G}$. The set of all closed projections is denoted by $P'(\mathscr{G})$. $P'(\mathscr{G})$ has a partial order induced from $P(\mathscr{G})$.

Lemma (37.7). *Let e and f be projections of a Baer *-semigroup $\mathscr{G}$. If $ef=fe$ then $ef=e\wedge f$ in $P(\mathscr{G})$. If moreover e and f are closed then ef is closed and $ef=e\wedge f$ in $P'(\mathscr{G})$.*

Proof. Evidently ef is a projection such that $ef\leqq e$ and $ef\leqq f$. If g is a projection of $\mathscr{G}$ such that $g\leqq e$ and $g\leqq f$, then we have $g=eg=efg$, whence $g\leqq ef$. Hence $ef=e\wedge f$ in $P(\mathscr{G})$.

If e and f are closed, then by (37.5.3) and (37.5.4) we have $(ef)''=(fe)''\leqq e''=e$ and $(ef)''\leqq f''=f$, whence $(ef)''\leqq ef$. Since $ef\leqq(ef)''$ by (37.5.5), ef is closed, and hence $ef=e\wedge f$ in $P'(\mathscr{G})$. ▯

Theorem (37.8). *Let $\mathcal{G}$ be a Baer *-semigroup. The set $P'(\mathcal{G})$ of closed projections of $\mathcal{G}$ forms an orthomodular lattice with the orthocomplementation $e \to e'$, and in $P'(\mathcal{G})$*

$$e \wedge f = e(f'e)' = (f'e)'e. \tag{37.8.1}$$

Proof. (I) Let $e, f \in P'(\mathcal{G})$ and put $a = f'e$. We shall show that the meet $e \wedge f$ in $P'(\mathcal{G})$ exists and (37.8.1) holds. We have $e' \leqq a'$ by (37.5.3), and hence e' commutes with a'. Since $e = e''$ commutes with a' by (37.5.8), we have $ea' = e \wedge a'$ in $P'(\mathcal{G})$ by (37.7). Moreover, since $f'ea' = aa' = 0$, we have $ea' \leqq f'' = f$ by (37.5.2). Hence ea' is a lower bound of $\{e, f\}$. If $g \in P'(\mathcal{G})$ is a lower bound of $\{e, f\}$, then $ag = f'eg = f'g = f'fg = 0$, whence $g \leqq a'$. Hence $g \leqq e \wedge a' = ea'$. Therefore $ea' = e \wedge f$ in $P'(\mathcal{G})$.

(II) Since the mapping $e \to e'$ is an involutorial dual-automorphism of $P'(\mathcal{G})$, the element $(e' \wedge f')'$ is the join $e \vee f$ in $P'(\mathcal{G})$. Hence $P'(\mathcal{G})$ forms a lattice. Moreover $e \wedge e' = 0$ for every $e \in P'(\mathcal{G})$, since $ee' = e'e = 0$. Hence $e \to e'$ is an orthocomplementation in $P'(\mathcal{G})$.

(III) For any $e, f \in P'(\mathcal{G})$, since $e' \wedge f' = e'(fe')' = e' \wedge (fe')'$ by (37.8.1), we have

$$e \vee f = e \vee (fe')''. \tag{1}$$

If $e \leqq f$, then since f commutes with e' by (37.5.8), we have $(fe')'' = fe' = f \wedge e'$. Hence by (1)

$$f = e \vee f = e \vee (f \wedge e').$$

By (29.13), $P'(\mathcal{G})$ is orthomodular. □

Remark (37.9). It can be proved that any orthomodular lattice is isomorphic to the lattice $P'(\mathcal{G})$ of closed projections of some Baer *-semigroup (Foulis [1]).

Lemma (37.10). *Let $\mathcal{G}$ be a Baer *-semigroup. If $e, f \in P'(\mathcal{G})$, then*

$$(ef)'' = (e \vee f') \wedge f \quad \textit{in } P'(\mathcal{G}).$$

Proof. Since $f' \leqq (ef)'$ by (37.5.3), we have $(ef)'' \leqq f'' = f$. Hence by (37.8.1)

$$(ef)'' = f \wedge (ef)'' = f((ef)'f)'.$$

On the other hand, since f commutes with $f \wedge e'$, it commutes with $(f \wedge e')' = e \vee f'$. Hence by (37.7)

$$(e \vee f') \wedge f = f(f \wedge e')',$$

and moreover $f \wedge e' = (ef)'f$ by (37.8.1). Therefore

$$(e \vee f') \wedge f = f((ef)'f)' = (ef)''. \quad □$$

Theorem (37.11). *If a Baer *-semigroup $\mathscr{G}$ satisfies the following condition:*

(37.11.1) *For any $e, f \in P'(\mathscr{G})$ there exists a *-automorphism θ of $\mathscr{G}$ such that $\theta(ef) = fe$ and $\theta(fe) = ef$,*

then the lattice $P'(\mathscr{G})$ is O-symmetric.

Proof. It suffices to show that $P'(\mathscr{G})$ satisfies the condition (36.14.1). If $e \wedge f = 0$ and $e \vee f = 1$ in $P'(\mathscr{G})$, then by (37.10)

(1) $$(ef')'' = (e \vee f) \wedge f' = f' \quad \text{and} \quad (f'e)'' = (f' \vee e') \wedge e = e.$$

By (37.11.1), there exists a *-automorphism θ of $\mathscr{G}$ such that $\theta(ef') = f'e$ and $\theta(f'e) = ef'$. It is easy to show that the restriction of θ on $P'(\mathscr{G})$ is an automorphism of $P'(\mathscr{G})$. Moreover, since $\theta(x') = \theta(x)'$ for every $x \in \mathscr{G}$, by (1) we have

$$\theta(e) = \theta((f'e)'') = (\theta(f'e))'' = (ef')'' = f' \quad \text{and}$$
$$\theta(f) = \theta((ef')') = (\theta(ef'))' = (f'e)' = e'.$$

Therefore $P'(\mathscr{G})$ satisfies (36.14.1). □

Definition (37.12). A **-ring* $\mathscr{A}$ is a ring with an involutorial anti-automorphism $x \to x^*$ ($(x+y)^* = x^* + y^*$, $(xy)^* = y^* x^*$ and $x^{**} = x$). Hence it is a *-semigroup. A *-ring $\mathscr{A}$ is called a *Baer *-ring* when under its multiplication it is a Baer *-semigroup. A Baer *-ring has the two-sided unit element by (37.4).

Theorem (37.13). *Let $\mathscr{A}$ be a Baer *-ring. Every projection e of $\mathscr{A}$ is closed and $e' = 1 - e$. The set $P(\mathscr{A})$ of all projections of $\mathscr{A}$ forms an orthomodular lattice with an orthocomplementation $e \to 1 - e$.*

Proof. Let $e \in P(\mathscr{A})$. Evidently $1 - e \in P(\mathscr{A})$. We have $e' = 1 - e$, because $ex = 0$ implies $x = (1-e)x \in (1-e)\mathscr{A}$ and conversely $x \in (1-e)\mathscr{A}$ implies $ex = 0$. Moreover e is closed since $e'' = 1 - (1-e) = e$. It follows from (37.8) that $P(\mathscr{A})$ $(= P'(\mathscr{A}))$ forms an orthomodular lattice where $e \to 1 - e$ is an orthocomplementation. □

Theorem (37.14). *If a Baer *-ring $\mathscr{A}$ satisfies the following condition:*

(SR) *For any $a \in \mathscr{A}$ there exists an element $|a| \in \mathscr{A}$ such that $|a|^* = |a|$, $|a|^2 = a^* a$ and $|a|$ commutes with every element that commutes with $a^* a$* ($|a|$ is a square root of $a^* a$),

then the lattice $P(\mathscr{A})$ is O-symmetric.

Proof. It suffices to show that $\mathscr{A}$ satisfies the condition (37.11.1).

Let $e, f \in P'(\mathscr{A}) = P(\mathscr{A})$. We put $a = e + f - 1$. Then $a^* = a$, and since $a^2 = ef + fe - e - f + 1$, we have

$$ea^2 = efe = a^2 e \quad \text{and} \quad fa^2 = fef = a^2 f.$$

By (SR) there exists $|a| \in \mathscr{A}$ such that

$$|a|^* = |a|, \qquad |a|^2 = a^2, \qquad |a|e = e|a| \quad \text{and} \quad |a|f = f|a|.$$

We put $a^+ = |a| + a$ and $a^- = |a| - a$. Since

$$|a|a = |a|(e + f - 1) = (e + f - 1)|a| = a|a|,$$

we have $a^- a^+ = |a|^2 - a^2 = 0$. By (37.5.7) we have $(a^-)'' a^+ = 0$, whence $a^+(a^-)'' = ((a^-)'' a^+)^* = 0$. Hence by (37.5.5) we have

(1) $$2|a|(a^-)'' = (a^+ + a^-)(a^-)'' = a^-(a^-)'' = a^-.$$

Now we put $s = 1 - 2(a^-)''$. Then since $s^* = s$ and $s^2 = 1 - 4(a^-)'' + 4(a^-)'' = 1$, the mapping $x \to sxs$ is a *-automorphism of $\mathscr{A}$. By (1) we have

(2) $$|a|s = |a| - 2|a|(a^-)'' = |a| - a^- = a.$$

Moreover, since $s|a| = (|a|s)^* = a^* = a$, we have

(3) $$sa = s^2|a| = |a|.$$

By (2) and (3) we have

$$\begin{aligned} sefs &= s(a - f + 1)fs = safs = |a|fs = f|a|s = fa \\ &= f(e + f - 1) = fe \quad \text{and} \\ sfes &= s^2 efs^2 = ef. \end{aligned}$$

Therefore $\mathscr{A}$ satisfies (37.11.1). □

Remark (37.15). Any von Neumann algebra is a Baer *-ring satisfying the condition (SR). More generally, if a C*-algebra $\mathscr{A}$ is a Baer *-semigroup (in this case $\mathscr{A}$ is called a B_p^*-algebra in Rickart [1]), then it can be proved that $\mathscr{A}$ satisfies (SR), using the representation of a commutative C*-algebra by the *-algebra of continuous functions on a compact space.

Remark (37.16). A Baer *-semigroup $\mathscr{G}$ is called complete when for any non-empty subset S of $\mathscr{G}$ there exists a projection e of $\mathscr{G}$ such that

$$\{x \in \mathscr{G}\,;\, ax = 0 \text{ for every } a \in S\} = e\mathscr{G}.$$

It is easy to show that a Baer *-semigroup $\mathscr{G}$ is complete if and only if the lattice $P'(\mathscr{G})$ is complete.

It is proved in Kaplansky [1] that if a complete Baer *-ring $\mathscr{A}$ satisfies the condition (SR) and one more condition which is denoted by (EP) in Kaplansky [1] (in particular, if $\mathscr{A}$ is an AW*-algebra), then $P(\mathscr{A})$ is a dimension lattice.

38. Modular Pairs in Lattices of Projections

Lemma (38.1). *For two elements a and b of a Baer *-semigroup $\mathscr{G}$,*

$$(a''b)' = (ab)'.$$

Proof. It follows from (37.5.7) that $abx=0$ if and only if $a''bx=0$. Hence $(ab)'=(a''b)'$. □

Theorem (38.2). *Let $\mathscr{G}$ be a Baer *-semigroup and let, $e,f\in P'(\mathscr{G})$. Then, $(e,f)M$ in $P'(\mathscr{G})$ if and only if the element $a=e'f$ satisfies the following condition:*

(38.2.1) $$((ga^*)'a)'' = (a'g')'g' \quad \text{for every } g\in P'(\mathscr{G}).$$

Proof. (I) Let $g\in P'(\mathscr{G})$ and put $h=(gf)''$. Since $h=(g\vee f')\wedge f$ by (37.10), we have

(1) $$h'\wedge f = ((g'\wedge f)\vee f')\wedge f = g'\wedge f,$$

since $(f',f)M$. By (38.1)

$$(g(e'f)^*)' = (gfe')' = (he')',$$

whence by (37.8.1)

$$(g(e'f)^*)'e' = (he')'e' = e'\wedge h'.$$

Hence by (37.10) we have

(2) $$((g(e'f)^*)'e'f)'' = ((h'\wedge e')\vee f')\wedge f.$$

On the other hand, by (37.8.1), (37.10) and (1), we have

(3) $$((e'f)'g')'g' = (e'f)''\wedge g' = ((e'\vee f')\wedge f)\wedge g' = h'\wedge(e'\vee f')\wedge f.$$

(II) If

(4) $$((h_1'\wedge e')\vee f')\wedge f = h_1'\wedge(e'\vee f')\wedge f \quad \text{for every } h_1\leqq f,$$

then it follows from (2) and (3) that $a=e'f$ satisfies (38.2.1). Conversely, if $a=e'f$ satisfies (38.2.1) then (4) holds, because if we put $g=h_1$ in (I) then $h=(h_1f)''=h_1''=h_1$. Hence it suffices to prove that $(e,f)M$ is equivalent to (4).

Taking orthocomplements, (4) is equivalent to

(5) $((h_1 \vee e) \wedge f) \vee f' = h_1 \vee (e \wedge f) \vee f'$ for every $h_1 \leqq f$.

If $(e,f)M$ then evidently (5) holds. If (5) holds, then taking the meet of each side of (5) with f and using $(f',f)M$, we have

$$(h_1 \vee e) \wedge f = h_1 \vee (e \wedge f) \quad \text{for every } h_1 \leqq f.$$

Hence $(e,f)M$. □

Remark (38.3). The theorem (37.11) can be proved by (38.2), replacing "$\theta(ef) = fe$ and $\theta(fe) = ef$" by only "$\theta(ef) = fe$". In fact, since θ is a *-automorphism, an element a satisfies the condition (38.2.1) if and only if $\theta(a)$ satisfies the same condition. Hence, if $\theta(e'f) = fe'$, then by (38.2) $(e,f)M$ implies $(f', e')M$.

Definition (38.4). Let $\mathscr{G}$ be a Baer *-semigroup. An element a of $\mathscr{G}$ is called *range-closed* when

(38.4.1) $g \in P'(\mathscr{G})$, $g \leqq a''$ and $(g a^*)'' = (a^*)''$ together imply $g = a''$.

If $\mathscr{G}$ is the Baer *-semigroup of all bounded linear operators of a Hilbert space H, then it can be proved that an operator $a \in \mathscr{G}$ satisfies (38.4.1) if and only if the range of a is closed in H (see Foulis [2]).

Next we shall show that (38.2.1) and (38.4.1) are equivalent.

Definition (38.5). Let $\mathscr{G}$ be a Baer *-semigroup. For each element a of $\mathscr{G}$ we define a mapping φ_a of $P'(\mathscr{G})$ into itself by

(38.5.1) $$\varphi_a(e) = (e a)''.$$

Evidently $\varphi_a(0) = 0$ and $\varphi_a(1) = a''$ for every $a \in \mathscr{G}$. If $f \in P'(\mathscr{G})$ then by (37.10)

$$\varphi_f(e) = (e \vee f') \wedge f.$$

Lemma (38.6). *Let a be an element of a Baer *-semigroup $\mathscr{G}$ and let $e,f \in P'(\mathscr{G})$.*

(38.6.1) $$\varphi_{a^*}(a') = 0.$$

(38.6.2) $$e \leqq f \quad \textit{implies} \quad \varphi_a(e) \leqq \varphi_a(f).$$

(38.6.3) $$\varphi_a(\varphi_{a^*}(e)') \leqq e' \wedge a''.$$

(38.6.4) $$\varphi_a(e \vee f) = \varphi_a(e) \vee \varphi_a(f).$$

(38.6.5) $$(\varphi_a \varphi_{a^*}(e)) \vee (e' \wedge a'') = a''.$$

Proof. (I) $\varphi_{a^*}(a') = (a' a^*)'' = 0$ by (37.5.1).

(II) If $e \leqq f$, then by (37.5.3) we have

$$(f a)' \leqq (e f a)' = (e a)',$$

whence $(e a)'' \leqq (f a)''$ by (37.5.4).

(III) By (37.5.1) we have $(ea^*)' ae = (ea^*)'(ea^*)^* = 0$, whence $e \leqq ((ea^*)' a)'$ by (37.5.2). Therefore

$$e' \geqq ((ea^*)' a)'' = \varphi_a(\varphi_{a^*}(e)').$$

Moreover, by (II), $\varphi_a(\varphi_{a^*}(e)') \leqq \varphi_a(1) = a''$.

(IV) We put $g = \varphi_a(e) \vee \varphi_a(f)$. By (II) we have $g \leqq \varphi_a(e \vee f)$. On the other hand, since $g' \leqq \varphi_a(e)'$, by (II) and (III) we have

$$\varphi_{a^*}(g') \leqq \varphi_{a^*}(\varphi_a(e)') \leqq e',$$

whence $e \leqq \varphi_{a^*}(g')'$. Similarly $f \leqq \varphi_{a^*}(g')'$. Hence, by (II) and (III) we have

$$\varphi_a(e \vee f) \leqq \varphi_a(\varphi_{a^*}(g')') \leqq g'' = g.$$

(V) We put $g = e \vee a'$ and $h = (\varphi_a \varphi_{a^*}(e))' \wedge g$. We shall show $a'' = h'$. Since $\varphi_a \varphi_{a^*}(e) \leqq \varphi_a(1) = a''$ and $g' = e' \wedge a'' \leqq a''$, we have $h' = \varphi_a \varphi_{a^*}(e) \vee g' \leqq a''$. On the other hand, by (III) we have

$$\varphi_{a^*}((\varphi_a \varphi_{a^*}(e))') \leqq \varphi_{a^*}(e)' \wedge (a^*)'' \leqq \varphi_{a^*}(e)',$$

whence $\varphi_{a^*}(h) \leqq \varphi_{a^*}(e)' \wedge \varphi_{a^*}(g)$. This implies $\varphi_{a^*}(h) = 0$, since $\varphi_{a^*}(g) = \varphi_{a^*}(e) \vee \varphi_{a^*}(a') = \varphi_{a^*}(e)$ by (IV) and (I). Hence by (III) we have

$$h' \geqq \varphi_a(\varphi_{a^*}(h)') = \varphi_a(1) = a''.$$

Therefore, $a'' = h' = (\varphi_a \varphi_{a^*}(e)) \vee (e' \wedge a'')$. □

Theorem (38.7). *Let $\mathscr{G}$ be a Baer *-semigroup and let $e, f \in P'(\mathscr{G})$. Then, $(e,f)M$ in $P'(\mathscr{G})$ if and only if the element $e'f$ is range-closed.*

Proof. By (38.2), it suffices to show that an element $a \in \mathscr{G}$ is range-closed if and only if a satisfies (38.2.1).

Assume that a satisfies (38.2.1). Let $g \in P'(\mathscr{G})$, $g \leqq a''$ and $(ga^*)'' = (a^*)''$. By (37.8.1) and (37.5.1) we have

$$a'' - g = g' \wedge a'' = (a'g')'g' = ((ga^*)' a)'' = ((a^*)' a)'' = 0.$$

Hence, a is range-closed.

Assume that a is range-closed, and let g be an arbitrary element of $P'(\mathscr{G})$. We remark that the left side of the equation in (38.2.1) is $\varphi_a(\varphi_{a^*}(g)')$ and the right side is $g' \wedge a''$. Since $\varphi_a(\varphi_{a^*}(g)') \leqq g' \wedge a''$ by (38.6.3), it suffices to show that $(g' \wedge a'') - \varphi_a(\varphi_{a^*}(g)') = 0$. We put $e = \varphi_{a^*}(g)'$ and $f = \varphi_a(e) \vee g \vee a'$. Since $e' = \varphi_{a^*}(g) \leqq \varphi_{a^*}(1) = (a^*)''$, by (38.6.5) we have

$$(a^*)'' = (\varphi_{a^*} \varphi_a(e)) \vee (e' \wedge (a^*)'') = (\varphi_{a^*} \varphi_a(e)) \vee e'.$$

Hence, by (38.6.4) and (38.6.1) we have

$$\varphi_{a^*}(f) = (\varphi_{a^*} \varphi_a(e)) \vee \varphi_{a^*}(g) \vee \varphi_{a^*}(a') = (\varphi_{a^*} \varphi_a(e)) \vee e' = (a^*)''.$$

Since $a' \leqq f$, we have $f = a' \vee (f \wedge a'')$ by the orthomodularity. Hence

(1) $$(a^*)'' = \varphi_{a^*}(f) = \varphi_{a^*}(a') \vee \varphi_{a^*}(f \wedge a'') = \varphi_{a^*}(f \wedge a'').$$

Since a is range-closed, (1) implies $f \wedge a'' = a''$. Hence $f \geqq a'' \vee a' = 1$. Therefore

$$(g' \wedge a'') - \varphi_a(\varphi_{a^*}(g)') = g' \wedge a'' \wedge \varphi_a(e)' = f' = 0. \quad \square$$

Finally, we add some remarks on commutativity in orthomodular lattices of projections.

Remark (38.8). Let $\mathscr{G}$ be a Baer *-semigroup and let $e, f \in P'(\mathscr{G})$. Then $e\,C\,f$ in $P'(\mathscr{G})$ if and only if $ef = fe$.

Proof. If $ef = fe$, then by (37.7) and (37.10) we have

$$e \wedge f = fe = (fe)'' = (f \vee e') \wedge e.$$

Hence, $e\,C\,f$ by (36.6.2).

Conversely, if $e\,C\,f$, then by (37.10) and (36.6.2) we have

$$(fe)'' = (f \vee e') \wedge e = e \wedge f \leqq f.$$

Hence by (37.5.5)

$$fe = fe(fe)'' = fe(fe)''f = fef.$$

Therefore we have $ef = (fe)^* = (fef)^* = fef = fe$. $\square$

Remark (38.9). Let H be a Hilbert space and let $\mathscr{B}(H)$ be the *-algebra of all bounded linear operators on H. The set $P(\mathscr{B}(H))$ of all projection operators forms a complete orthomodular lattice isomorphic to the lattice $L_c(H)$ of closed subspaces of H.

For any subset $\mathscr{S}$ of $\mathscr{B}(H)$ we put

$$\mathscr{S}' = \{T \in \mathscr{B}(H);\ TS = ST \text{ for every } S \in \mathscr{S}\}.$$

A *von Neumann algebra* $\mathscr{A}$ is a *-subalgebra of $\mathscr{B}(H)$ such that $\mathscr{A}'' = \mathscr{A}$ (see Dixmier [1]). It can be proved that if $\mathscr{A}$ is a von Neumann algebra then $\mathscr{A}$ is generated by the set $P(\mathscr{A})$ of its projection operators in the following sense: $\mathscr{A} = P(\mathscr{A})''$.

Now we shall show that if $\mathscr{A}$ is a von Neumann algebra then $P(\mathscr{A})$ is a C-closed sublattice of $P(\mathscr{B}(H))$. Put $\mathscr{S} = P(\mathscr{A}')$. Since $\mathscr{A}' = P(\mathscr{A})''' = P(\mathscr{A})'$, by (38.8) we have

$$C(P(\mathscr{A})) = P(\mathscr{B}(H)) \cap P(\mathscr{A})' = P(\mathscr{B}(H)) \cap \mathscr{A}' = \mathscr{S}.$$

Moreover, since $\mathscr{A}'$ is a von Neumann algebra, we have $\mathscr{S}'' = P(\mathscr{A}')'' = \mathscr{A}'$, whence $\mathscr{S}' = \mathscr{A}'' = \mathscr{A}$. Hence

$$P(\mathscr{A}) = P(\mathscr{S}') = C(\mathscr{S}) = CC(P(\mathscr{A})).$$

Therefore $P(\mathscr{A})$ is a C-closed sublattice of $P(\mathscr{B}(H))$.

PROBLEM 9. Is the completion by cuts of a relatively complemented lattice with 0 and 1 satisfying the condition (J) M-symmetric?

PROBLEM 10. Is there an orthomodular lattice satisfying (J) which is not O-symmetric?

PROBLEM 11. Is a C-closed sublattice of an orthomodular AC-lattice M-symmetric? Is a C-closed sublattice of an orthomodular O-symmetric AC-lattice O-symmetric?

References for Chapter VIII

For Section 35: A. Ramsay [1], M. F. Janowitz [2], S. Maeda [2], M. D. MacLaren [2], L. H. Loomis [1], R. J. Greechie [1].

For Section 36: M. Nakamura [1], D. J. Foulis [3], S. Maeda [4], S. S. Holland, Jr. [1] and [2], E. A. Schreiner [1].

For Section 37: D. J. Foulis [1] and [4], D. M. Topping [1], C. E. Richart [1], I. Kaplansky [1]; the proof of (37.11) is due to S. S. Holland, Jr.

For Section 38: D. J. Foulis [2] and [4], J. Dixmier [1].

Supplement

Recently, Problem 2 in p. 54 has been solved affirmatively by M. F. Janowitz. Here, by his idea, we shall show *the existence of a complete finite-modular AC-lattice which is not $\perp$-symmetric.*

Let Λ be a modular matroid lattice of infinite length having the lattice operations $\sqcup$ and $\sqcap$. Then there exists an infinite semi-orthogonal family of atoms p_n $(n=1,2,\dots)$ in Λ. Putting

$$a=p_1\sqcup p_3\sqcup\cdots \quad \text{and} \quad b=p_2\sqcup p_4\sqcup\cdots,$$

we have $a\notin\mathscr{J}(\Lambda)$, $b\notin\mathscr{J}(\Lambda)$ and moreover $a\sqcap b=0$ by (3.5) and (2.16). Let

$$L=\{u\sqcup v;\ u\in\Lambda[0,a]\cup\{b,1\} \text{ and } v\in\mathscr{J}(\Lambda)\}.$$

Evidently L satisfies the conditions (15.15.1) and (15.15.3). We shall show that

(1) $\qquad b\leqq x\leqq b\sqcup v$ with $v\in\mathscr{J}(\Lambda)$ implies $x\in L$, and

(2) $\qquad x\leqq a\sqcup v$ with $v\in\mathscr{J}(\Lambda)$ implies $x\in L$.

If $b\leqq x\leqq b\sqcup v$, then by the modularity of Λ we have $x=(b\sqcup v)\sqcap x =b\sqcup(v\sqcap x)$. Since $v\sqcap x\in\mathscr{J}(\Lambda)$, we have $x\in L$. If $x\leqq a\sqcup v$, then since Λ is relatively complemented, there exists a complement v_1 of $x\sqcap a$ in $\Lambda[0,x]$. Then v_1 is subperspective to v, since $v_1\leqq a\sqcup v$ and $v_1\sqcap a =v_1\sqcap x\sqcap a=0$. Hence $v_1\in\mathscr{J}(\Lambda)$ by (11.9). Since $x=(x\sqcap a)\sqcup v_1$, we have $x\in L$. Thus (1) and (2) has been proved. If $x_\alpha\in L$ for every α then it is easy to show that $\bigsqcap_\alpha x_\alpha\in L$ by (1) and (2). Hence L satisfies (15.15.2), and then L forms a complete AC-lattice. Moreover, L is finite-modular by (15.15.7).

Evidently $a\wedge b=0$ in L. We shall show that the pair (a,b) is modular but (b,a) is not. We need to show that

(3) $\qquad b_1<b$ with $b_1\in L$ implies $b_1\in\mathscr{J}(\Lambda)$, and

(4) $\qquad a_1\leqq a$ with $a_1\notin\mathscr{J}(\Lambda)$ implies $a_1\vee b=1$ in L.

To prove (3), we put $b_1=u\sqcup v$ with $u\in\Lambda[0,a]\cup\{b,1\}$ and $v\in\mathscr{J}(\Lambda)$. Since $u\in\{b,1\}$ contradicts $b_1<b$, we have $u\leqq a$, whence $u\leqq a\sqcap b_1 \leqq a\sqcap b=0$. Hence $b_1=v\in\mathscr{J}(\Lambda)$. To prove (4), let $u\sqcup v$ be an upper

bound of $\{a_1, b\}$. If we had $u \leqq a$, then $b \leqq u \sqcup v \leqq a \sqcup v$. Since $b \sqcap a = 0$, b is subperspective to v in Λ, whence $b \in \mathscr{J}(\Lambda)$, a contradiction. If we had $u = b$, then $a_1 \leqq b \sqcup v$ and then a_1 is subperspective to v, whence $a_1 \in \mathscr{J}(\Lambda)$, a contradiction. Hence $u = 1$, which means that $a_1 \vee b = 1$.

Now, if $b_1 < b$ in L, then since $b_1 \in \mathscr{J}(\Lambda)$ by (3), we have $b_1 \sqcup a \in L$. By (15.15.4) and (15.15.5) and by the modularity of Λ,

$$(b_1 \vee a) \wedge b = (b_1 \sqcup a) \sqcap b = b_1 \sqcup (a \sqcap b) = b_1 = b_1 \vee (a \wedge b).$$

Hence $(a, b) M$. On the other hand, putting $a_1 = p_3 \sqcup p_5 \sqcup \cdots$, we have $a_1 < a$ and $a_1 \notin \mathscr{J}(\Lambda)$. Since $a_1 \vee b = 1$ by (4), we have

$$(a_1 \vee b) \wedge a = a > a_1 = a_1 \vee (b \wedge a).$$

Hence $(b, a) \overline{M}$. Thus L is not $\perp$-symmetric.

Problem. Is there a Hausdorff topological vector space E such that $L_c(E)$ is not M-symmetric? (Cf. (31.4).)

Bibliography

AMEMIYA, I. AND H. ARAKI

[1] A remark on Piron's paper. Publications Research Inst. Math. Sci., Kyoto Univ., Ser. A **2**, 423—427 (1966).

BAER, R.

[1] Linear algebra and projective geometry. New York: Academic Press 1952.

BIRKHOFF, G.

[1] Lattice theory. Third edition. New York: Amer. Math. Soc. Colloq. Publ. 1967.

BIRKHOFF, G. AND J. VON NEUMANN

[1] The logic of quantum mechanics. Ann. of Math. **37**, 823—843 (1936).

DIXMIER, J.

[1] Les algèbres d'opérateurs dans l'espace hilbertien. Paris: Gauthier-Villars 1957.

DUBREIL-JACOTIN, M. L., L. LEISIEUR AND R. CROISOT

[1] Leçons sur la théorie des treillis des structures algébriques ordonnées et des treillis géométriques. Paris: Gauthier-Villars 1953.

FOULIS, D. J.

[1] Baer *-semigroups. Proc. Amer. Math. Soc. **11**, 648—654 (1960).

[2] Conditions for the modularity of an orthomodular lattice. Pacific J. Math. **11**, 889—895 (1961).

[3] A note on orthomodular lattices. Portugal Math. **21**, 65—72 (1962).

[4] Relative inverse in Baer *-semigroups. Michigan Math. J. **10**, 65—84 (1963).

GREECHIE, R. J.

[1] On the structure of orthomodular lattices satisfying the chain condition. J. Combinatorial Theory **4**, 210—218 (1968).

HALMOS, P. R.

[1] Introduction to Hilbert space and theory of spectral multiplicity. New York: Van Nostrand 1957.

HOLLAND, JR., S. S.

[1] A Radon-Nikodym theorem in dimension lattices. Trans. Amer. Math. Soc. **108**, 66—87 (1963).

[2] Distributivity and perspectivity in orthomodular lattices. Trans. Amer. Math. Soc. **112**, 330—343 (1964).

[3] Partial solution to Mackey's problem about modular pairs and completeness. Canad. J. Math. **21**, 1518—1525 (1969).

Hsu, C.-J.
[1] On lattice theoretic characterization of the parallelism in affine geometry. Ann. of Math. **50,** 188—203 (1949).

Janowitz, M. F.
[1] A note on normal ideals. J. Sci. Hiroshima Univ., Ser. A-I **30,** 1—9 (1966).
[2] On conditionally continuous lattices II. Tech. Report Univ. New Mexico, No. 142, 1967.
[3] Note on a theorem of Zierler (unpublished).
[4] Section semicomplemented lattices. Math. Z. **108,** 63—76 (1968).
[5] On the modular relation in atomistic lattices. Fund. Math. (to appear).

Jónsson, B.
[1] Lattice-theoretic approach to projective and affine geometry. In: Proc. Internat. Sympos. Axiomatic Method, pp. 183—203. Amsterdam: North-Holland 1959.

Kaplansky, I.
[1] Rings of operators. New York: Benjamin 1968.

Loomis, L. H.
[1] The lattice theoretic background of the dimension theory of operator algebras. Memoirs of Amer. Math. Soc., No. 18, 1955.

Mackey, G. W.
[1] On infinite dimensional linear spaces. Trans. Amer. Math. Soc. **57,** 155—207 (1945).

MacLane, S.
[1] A lattice formulation for transcendence degrees and p-bases. Duke Math. J. **4,** 455—468 (1938).

MacLaren, M. D.
[1] Atomic orthocomplemented lattices. Pacific J. Math. **14,** 597—612 (1964).
[2] Nearly modular orthocomplemented lattices. Trans. Amer. Math. Soc. **114,** 401—416 (1965).

Maeda, F.
[1] Matroid lattices of infinite length. J. Sci. Hiroshima Univ., Ser. A **15,** 177—182 (1952).
[2] Kontinuierliche Geometrien. Berlin: Springer-Verlag 1958.
[3] Decomposition of general lattices into direct summands of types I, II and III. J. Sci. Hiroshima Univ., Ser. A **23,** 151—170 (1959).
[4] Modular centers of affine matroid lattices. J. Sci. Hiroshima Univ., Ser. A-I **27,** 73—84 (1963).
[5] Parallel mappings and comparability theorems in affine matroid lattices. J. Sci. Hiroshima Univ., Ser. A-I **27,** 85—96 (1963).
[6] Point-free parallelism in Wilcox lattices. J. Sci. Hiroshima Univ., Ser. A-I **28,** 10—32 (1964).
[7] Perspectivity of points in matroid lattices. J. Sci. Hiroshima Univ., Ser. A-I **28,** 101—112 (1964).

Maeda, S.
[1] On relatively semi-orthocomplemented lattices. J. Sci. Hiroshima Univ., Ser. A **24,** 155—161 (1960).

[2] Dimension theory on relatively semi-orthocomplemented complete lattice. J. Sci. Hiroshima Univ., Ser. A-I **25**, 369—404 (1961).
[3] On the symmetry of the modular relation in atomic lattices. J. Sci. Hiroshima Univ., Ser. A-I **29**, 165—170 (1965).
[4] On condition for orthomodularity. Proc. Japan Acad. **42**, 247—251 (1966).
[5] Infinite distributivity in complete lattices. Memoirs Ehime Univ., Sect. II, Ser. A **5**, 11—13 (1966).
[6] On atomistic lattices with the covering property. J. Sci. Hiroshima Univ., Ser. A-I **31**, 105—121 (1967).
[7] Modular pairs in atomistic lattices with the covering property. Proc. Japan. Acad. **45**, 149—153 (1969).
[8] Point-free parallelism and Wilcox lattices (to appear).

MARTINEAU, A.
[1] Sur une propriété caractéristique d'un produit de droites. Arch. Math. **11**, 423—426 (1960).

MCLAUGHLIN, J. E.
[1] The normal completion of a complemented modular point lattice. In: Proc. Sympos. Pure Math., Vol. 2, pp. 78—80. Amer. Math. Soc. 1961.

NAKAMURA, M.
[1] The permutability in a certain orthocomplemented lattice. Kôdai Math. Sem. Reports **9**, 158—160 (1957).

PIRON, C.
[1] Axiomatique quantique. Helv. Phys. Acta **37**, 439—468 (1964).

RAMSAY, A.
[1] Dimension theory in complete orthocomplemented weakly modular lattice. Trans. Amer. Math. Soc. **116**, 9—31 (1965).

RICKART, C. E.
[1] Banach algebras with an adjoint operation. Ann. of Math. **47**, 528—550 (1946).

SACHS, D.
[1] Partition and modulated lattices. Pacific J. Math. **11**, 325—345 (1961).

SASAKI, U.
[1] Lattice theoretic characterization of an affine geometry of arbitrary dimensions. J. Sci. Hiroshima Univ., Ser. A **16**, 223—238 (1953).
[2] Semi-modularity in relatively atomic, upper continuous lattices. J. Sci. Hiroshima Univ., Ser. A **16**, 409—416 (1953).
[3] Lattice theoretic characterization of geometries satisfying "Axiome der Verknüpfung". J. Sci. Hiroshima Univ., Ser. A **16**, 417—423 (1953).
[4] Orthocomplemented lattices, satisfying the exchange axiom. J. Sci. Hiroshima Univ., Ser. A **17**, 293—302 (1954).

SASAKI, U. AND S. FUJIWARA
[1] The decomposition of matroid lattices. J. Sci. Hiroshima Univ., Ser. A **15**, 183—188 (1952).

SCHAEFER, H. H.
[1] Topological vector spaces. New York: Macmillan 1966.

SCHREINER, E. A.
[1] Modular pairs in orthomodular lattices. Pacific J. Math. **19**, 519—528 (1966).

TOPPING, D. M.
[1] Asymptoticity and semimodularity in projection lattices. Pacific J. Math. **20,** 317—325 (1967).

WILCOX, L. R.
[1] Modularity in the theory of lattices. Ann. of Math. **40,** 490—505 (1939).
[2] A note on complementation in lattices. Bull. Amer. Math. Soc. **48,** 453—458 (1942).

WILLE, R.
[1] Halbkomplementäre Verbände. Math. Z. **94,** 1—31 (1966).

Subject Index

List of Special Symbols

Die Grundlehren der mathematischen Wissenschaften in Einzeldarstellungen mit besonderer Berücksichtigung der Anwendungsgebiete

---

2. Knopp: Theorie und Anwendung der unendlichen Reihen. DM 48,—; US $ 13.20
3. Hurwitz: Vorlesungen über allgemeine Funktionentheorie und elliptische Funktionen. DM 49,—; US $ 13.50
4. Madelung: Die mathematischen Hilfsmittel des Physikers. DM 49,70; US $ 13.70
10. Schouten: Ricci-Calculus. DM 58,60; US $ 16.20
14. Klein: Elementarmathematik vom höheren Standpunkte aus. 1. Band: Arithmetik, Algebra, Analysis. DM 24,—; US $ 6.60
15. Klein: Elementarmathematik vom höheren Standpunkte aus. 2. Band: Geometrie. DM 24,—; US $ 6.60
16. Klein: Elementarmathematik vom höheren Standpunkte aus. 3. Band: Präzisions- und Approximationsmathematik. DM 19,80; US $ 5.50
20. Pólya/Szegö: Aufgaben und Lehrsätze aus der Analysis II: Funktionentheorie, Nullstellen, Polynome, Determinanten, Zahlentheorie. DM 38,—; US $ 10.50
22. Klein: Vorlesungen über höhere Geometrie. DM 28,—; US $ 7.70
26. Klein: Vorlesungen über nicht-euklidische Geometrie. DM 24,—; US $ 6.60
27. Hilbert/Ackermann: Grundzüge der theoretischen Logik. DM 38,—; US $ 10.50
30. Lichtenstein: Grundlagen der Hydromechanik. DM 38,—; US $ 10.50
31. Kellogg: Foundations of Potential Theory. DM 32,- ; US $ 8.80
32. Reidemeister: Vorlesungen über Grundlagen der Geometrie. DM 18,—; US $ 5.00
38. Neumann: Mathematische Grundlagen der Quantenmechanik. DM 28,—; US $ 7.70
40. Hilbert/Bernays: Grundlagen der Mathematik I. DM 68,—; US $ 18.70
43. Neugebauer: Vorlesungen über Geschichte der antiken mathematischen Wissenschaften. 1. Band: Vorgriechische Mathematik. DM 48,—; US $ 13.20
50. Hilbert/Bernays: Grundlagen der Mathematik II. DM 68.—; US $ 18.70
52. Magnus/Oberhettinger/Soni: Formulas and Theorems for the Special Functions of Mathematical Physics. DM 66,—; US $ 16.50
57. Hamel: Theoretische Mechanik. DM 84,—; US $ 23.10
58. Blaschke/Reichardt: Einführung in die Differentialgeometrie. DM 24,—; US $ 6.60
59. Hasse: Vorlesungen über Zahlentheorie. DM 69,—; US $ 19.00
60. Collatz: The Numerical Treatment of Differential Equations. DM 78,—; US $ 19.50
61. Maak: Fastperiodische Funktionen. DM 38,—; US $ 10.50
62. Sauer: Anfangswertprobleme bei partiellen Differentialgleichungen. DM 41,—; US $ 11.30
64. Nevanlinna: Uniformisierung. DM 49,50; US $ 13.70
66. Bieberbach: Theorie der gewöhnlichen Differentialgleichungen. DM 58,50; US $ 16.20
68. Aumann: Reelle Funktionen. DM 68,—; US $ 18.70
69. Schmidt: Mathematische Gesetze der Logik I. DM 79,—; US $ 21.80
71. Meixner/Schäfke: Mathieusche Funktionen und Sphäroidfunktionen mit Anwendungen auf physikalische und technische Probleme. DM 52,60; US $ 14.50
73. Hermes: Einführung in die Verbandstheorie. DM 46,—; US $ 12.70
74. Boerner: Darstellungen von Gruppen. DM 58,—; US $ 16.00
75. Rado/Reichelderfer: Continuous Transformations in Analysis, with an Introduction to Algebraic Topology. DM 59,60; US $ 16.40
76. Tricomi: Vorlesungen über Orthogonalreihen. DM 68,—; US $ 18.70

77. Behnke/Sommer: Theorie der analytischen Funktionen einer komplexen Veränderlichen. DM 79,—; US $ 21.80
78. Lorenzen: Einführung in die operative Logik und Mathematik. DM 54,—; US $ 14.90
80. Pickert: Projektive Ebenen. DM 48,60; US $ 13.40
81. Schneider: Einführung in die transzendenten Zahlen. DM 24,80; US $ 6.90
82. Specht: Gruppentheorie. DM 69,60; US $ 19.20
84. Conforto: Abelsche Funktionen und algebraische Geometrie. DM 41,80; US $ 11.50
86. Richter: Wahrscheinlichkeitstheorie. DM 68,—; US $ 18,70
88. Müller: Grundprobleme der mathematischen Theorie elektromagnetischer Schwingungen. DM 52,80; US $ 14.60
89. Pfluger: Theorie der Riemannschen Flächen. DM 39,20; US $ 10.80
90. Oberhettinger: Tabellen zur Fourier-Transformation. DM 39,50; US $ 10.90
91. Prachar: Primzahlverteilung. DM 58,—; US $ 16.00
93. Hadwiger: Vorlesungen über Inhalt, Oberfläche und Isoperimetrie. DM 49,80; US $ 13.70
94. Funk: Variationsrechnung und ihre Anwendung in Physik und Technik. DM 120,—; US $ 33.00
95. Maeda: Kontinuierliche Geometrien. DM 39,—; US $ 10.80
97. Greub: Linear Algebra. DM 39,20; US $ 9.80
98. Saxer: Versicherungsmathematik. 2. Teil. DM 48,60; US $ 13.40
99. Cassels: An Introduction to the Geometry of Numbers. DM 69,—; US $ 19.00
100. Koppenfels/Stallmann: Praxis der konformen Abbildung. DM 69,—; US $ 19.00
101. Rund: The Differential Geometry of Finsler Spaces. DM 59,60; US $ 16.40
103. Schütte: Beweistheorie. DM 48,—; US $ 13.20
104. Chung: Markov Chains with Stationary Transition Probabilities. DM 56,—; US $ 14.00
105. Rinow: Die innere Geometrie der metrischen Räume. DM 83,—; US $ 22.90
106. Scholz/Hasenjaeger: Grundzüge der mathematischen Logik. DM 98,—; US $ 27.00
107. Köthe: Topologische lineare Räume I. DM 78,—; US $ 21.50
108. Dynkin: Die Grundlagen der Theorie der Markoffschen Prozesse. DM 33,80; US $ 9.30
110. Dinghas: Vorlesungen über Funktionentheorie. DM 69,—; US $ 19.00
112. Morgenstern/Szabó: Vorlesungen über theoretische Mechanik. DM 69,—; US $ 19.00
113. Meschkowski: Hilbertsche Räume mit Kernfunktion. DM 58,—; US $ 16.00
114. MacLane: Homology. DM 62,—; US $ 15.50
115. Hewitt/Ross: Abstract Harmonic Analysis. Vol. 1: Structure of Topological Groups, Integration Theory, Group Representations. DM 76,—; US $ 20.90
116. Hörmander: Linear Partial Differential Operators. DM 42,—; US $ 10.50
117. O'Meara: Introduction to Quadratic Forms. DM 48,—; US $ 13.20
118. Schäfke: Einführung in die Theorie der speziellen Funktionen der mathematischen Physik. DM 49,40; US $ 13.60
119. Harris: The Theory of Branching Processes. DM 36,—; US $ 9.90
120. Collatz: Funktionalanalysis und numerische Mathematik. DM 58,—; US $ 16.00
121.
122. Dynkin: Markov Processes. DM 96,—; US $ 26.40
123. Yosida: Functional Analysis. DM 66,—; US $ 16.50
124. Morgenstern: Einführung in die Wahrscheinlichkeitsrechnung und mathematische Statistik. DM 38,—; US $ 10.50
125. Itô/McKean: Diffusion Processes and Their Sample Paths. DM 58,—; US $ 16.00
126. Lehto/Virtanen: Quasikonforme Abbildungen. DM 38,—; US $ 10.50
127. Hermes: Enumerability, Decidability, Computability. DM 39,—; US $ 10.80

128. Braun/Koecher: Jordan-Algebren. DM 48,—; US $ 13.20
129. Nikodým: The Mathematical Apparatus for Quantum-Theories. DM 144,—; US $ 36.00
130. Morry: Multiple Integrals in the Calculus of Variations. DM 78,—; US $ 19.50
131. Hirzebruch: Topological Methods in Algebraic Geometry. DM 38,—; US $ 9.50
132. Kato: Perturbation Theory for Linear Operators. DM 79,20; US $ 19.80
133. Haupt/Künneth: Geometrische Ordnungen. DM 68,—; US $ 18.70
134. Huppert: Endliche Gruppen I. DM 156,—; US $ 42.90
135. Handbook for Automatic Computation. Vol 1/Part a: Rutishauser: Description of ALOGL 60. DM 58,—; US $ 14.50
136. Greub: Multilinear Algebra. DM 32,—; US $ 8.00
137. Handbook for Automatic Computation. Vol. 1/Part b: Grau/Hill/Langmaack: Translation of ALOGL 60. DM 64,—; US $ 16.00
138. Hahn: Stability of Motion. DM 72,—; US $ 19.80
139. Mathematische Hilfsmittel des Ingenieurs. Herausgeber: Sauer/Szabó. 1. Teil. DM 88,—; US $ 24.20
140. Mathematische Hilfsmittel des Ingenieurs. Herausgeber: Sauer/Szabó. 2. Teil. DM 136,—; US $ 37.40
141. Mathematische Hilfsmittel des Ingenieurs. Herausgeber: Sauer/Szabó. 3. Teil. DM 98,—; US $ 27.00
142. Mathematische Hilfsmittel des Ingenieurs. Herausgeber: Sauer/Szabó. 4. Teil. DM 124,—; US $ 34.10
143. Schur/Grunsky: Vorlesungen über Invariantentheorie. DM 32,—; US $ 8.80
144. Weil: Basic Number Theory. DM 48,—; US $ 12.00
145. Butzer/Berens: Semi-Groups of Operators and Approximation. DM 56,—; US $ 14.00
146. Treves: Locally Convex Spaces and Linear Differential Equations. DM 36,—; US $ 9.90
147. Lamotke: Semisimpliziale algebraische Topologie. DM 48,—; US $ 13.20
148. Chandrasekharan: Introduction to Analytic Number Theory. DM 28,—; US $ 7.00
149. Sario/Oikawa: Capacity Functions. DM 96,—; US $ 24.00
150. Iosifescu/Theodorescu: Random Processes and Learning. DM 68,—; US $ 18.70
151. Mandl: Analytical Treatment of One-dimensional Markov Processes. DM 36,—; US $ 9.80
152. Hewitt/Ross: Abstract Harmonic Analysis. Voll. II. Structure and Analysis for Compact Groups. Analysis on Locally Compact Abelian Groups. DM 140,—; US $ 38.50
153. Federer: Geometric Measure Theory. DM 118,—; US $ 29.50
154. Singer: Bases in Banach Spaces I. DM 112,—; US $ 30.80
155. Müller: Foundations of the Mathematical Theory of Electromagnetic Waves. DM 58,—; US $ 16.00
156. van der Waerden: Mathematical Statistics. DM 68,—; US $ 18.70
157. Prohorov/Rozanov: Probability Theory. DM 68,—; US $ 18.70
158. Constantinescu/Cornea: Potential Theory on Harmonic Spaces. In preparation
159. Köthe: Topological Vector Spaces I. DM 78,—; US $ 21.50
160. Agrest/Maksimov: Theory of Incomplete Cylindrical Functions and their Applications. In preparation
161. Bhatia/Szegö: Stability Theory of Dynamical Systems. DM 58,—; US $ 16.00
162. Nevanlinna: Analytic Functions. DM 76,—; US $ 21.00
163. Stoer/Witzgall: Convexity and Optimization in Finite Dimensions I. DM 54,—; US $ 14.90
164. Sario/Nakai: Classification Theory of Riemann Surfaces. DM 98,—; US $ 27.00
165. Mitrinović: Analytic Inequalities. DM 88,—; US $ 26.00

166. Grothendieck/Dieudonné: Eléments de Géométrie Algébrique I. DM 84,—; US $ 23.10
167. Chandrasekharan: Arithmetical Functions. DM 58,—; US $ 16.00
168. Palamodov: Linear Differential Operators with Constant Coefficients. DM 98,—; US $ 27.00
169. Rademacher: Topics in Analytic Number Theory. In preparation
170. Lions: Optimal Control of Systems Governed by Partial Differential Equations. DM 78,—; US $ 21.50
171. Singer: Best Approximation in Normed Linear Spaces by Elements of Linear Subspaces. DM 60,—; US $ 16.50
172. Bühlmann: Mathematical Methods in Risk Theory. DM 52,—; US $ 14.30
173. F. Maeda/S. Maeda: Theory of Symmetric Lattices. DM 48,—; US $ 13.20
174. Stiefel/Scheifele: Linear and Regular Celestial Mechanics. In preparation
175. Larsen: An Introduction to the Theory of Multipliers. In preparation
176. Grauert/Remmert: Analytische Stellenalgebren. In Vorbereitung
177. Flügge: Practical Quantum Mechanics (I. Teil: Kapitel I + II). In preparation
178. Flügge: Practical Quantum Mechanics (II Teil: Kapitel III—VII). In preparation